Apartment Blossom

楼房花朵

03

Apartment Blossom

楼房花朵

03

Apartment Blossom 楼房花朵 05

Apartment Blossom 楼房花朵 05

Apartment
Blossom

楼房花
朵

06

Apartment
Blossom

楼房花
朵

06

Apartment
Blossom
楼房花
朵
07

Apartment Blossom

楼房花朵

07

Apartment Blossom 楼房花朵 08

Apartment Blossom 楼房花朵 09

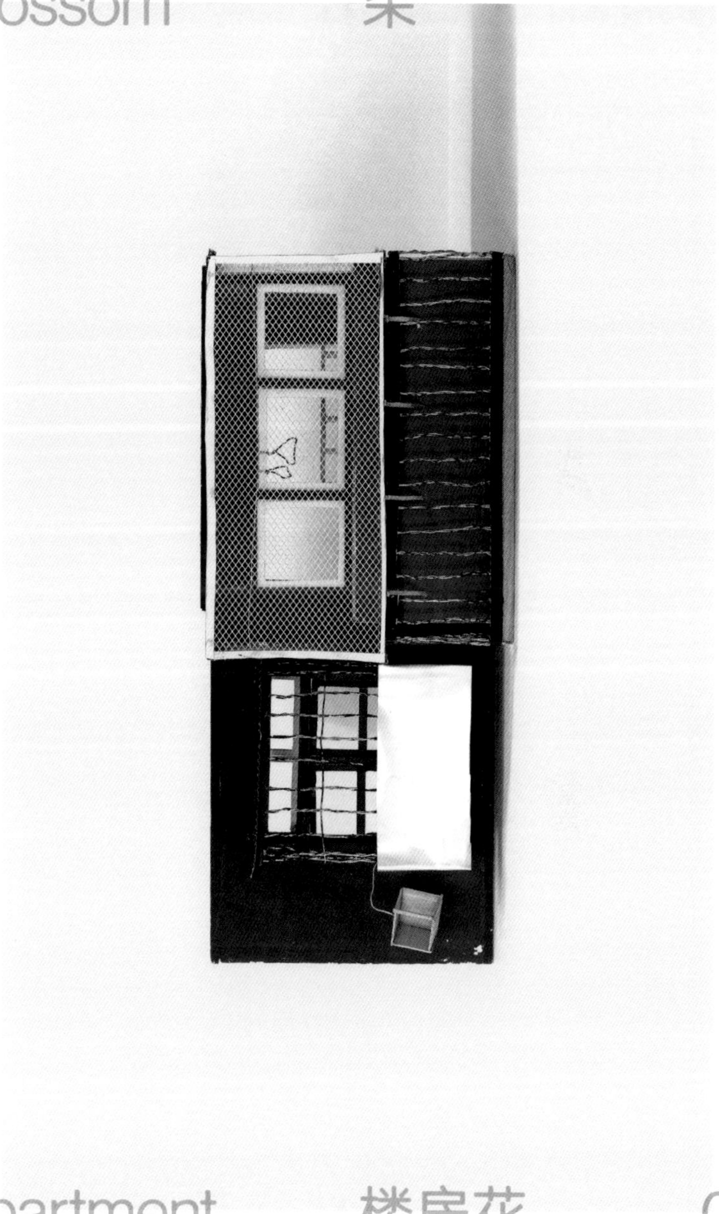

Apartment
Blossom
楼房花
朵
09

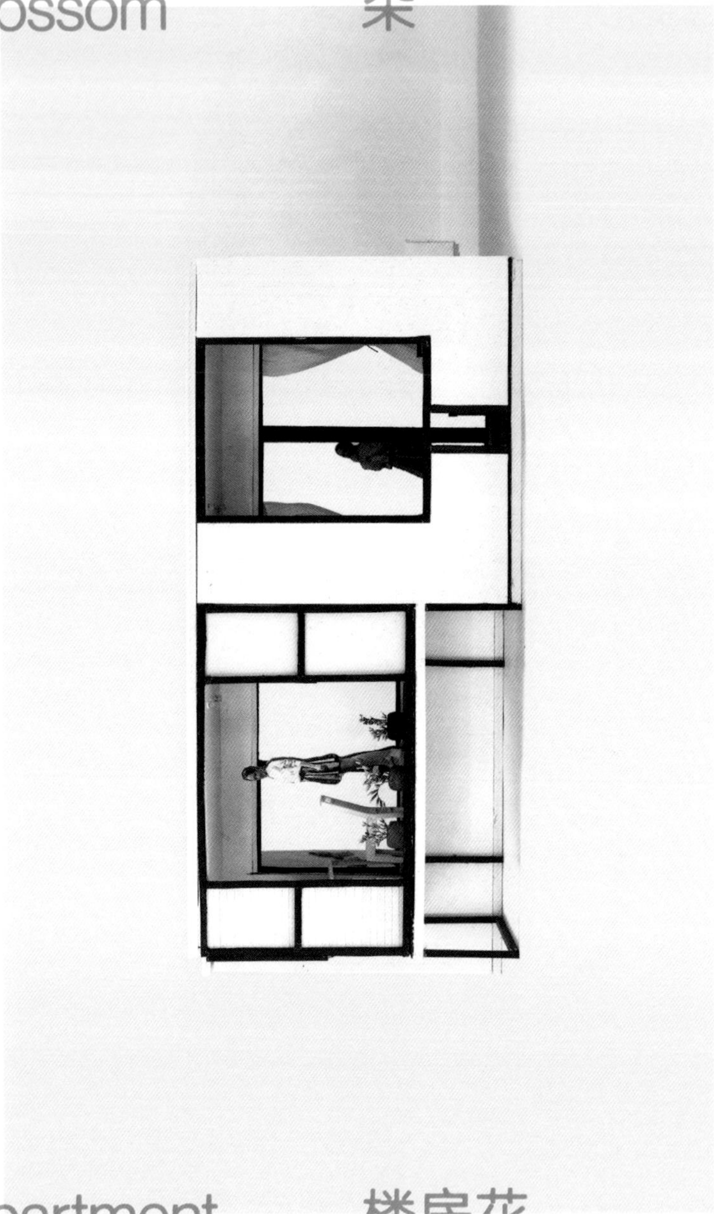

Apartment Blossom 楼房花朵 12

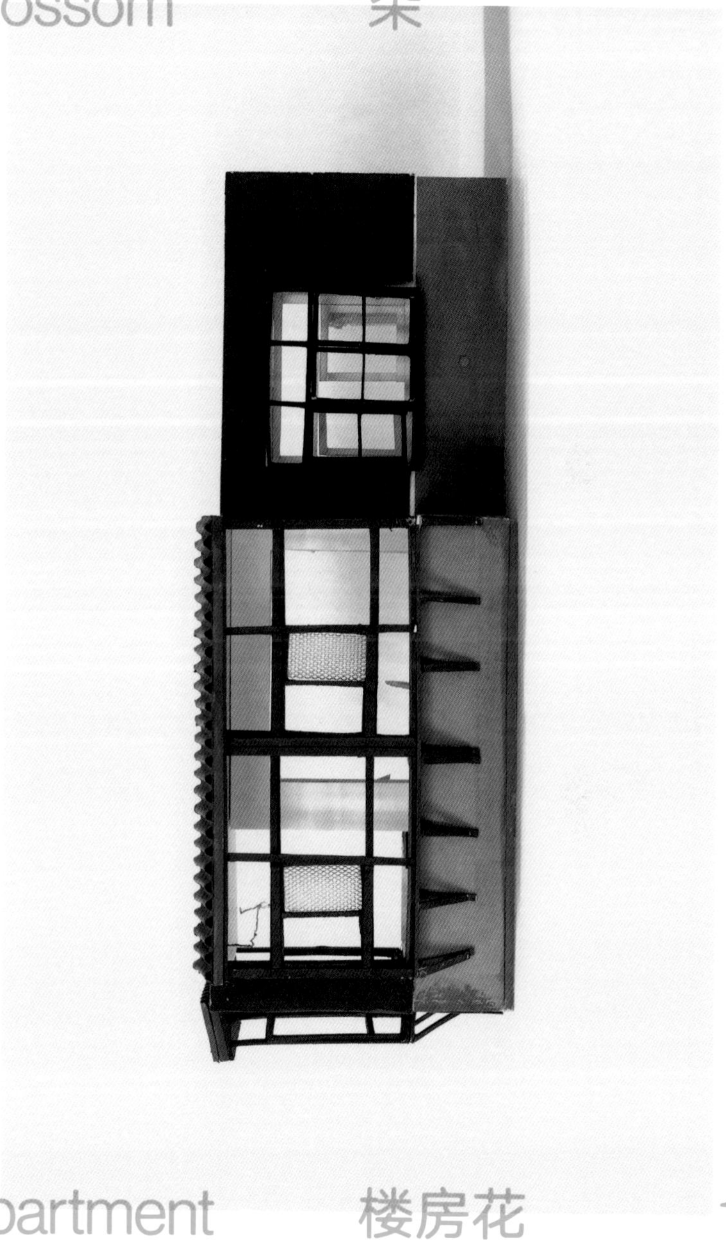

Apartment Blossom 楼房花朵 13

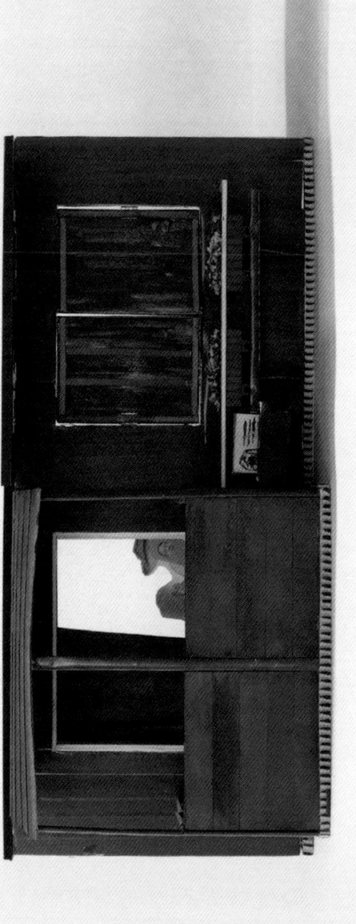

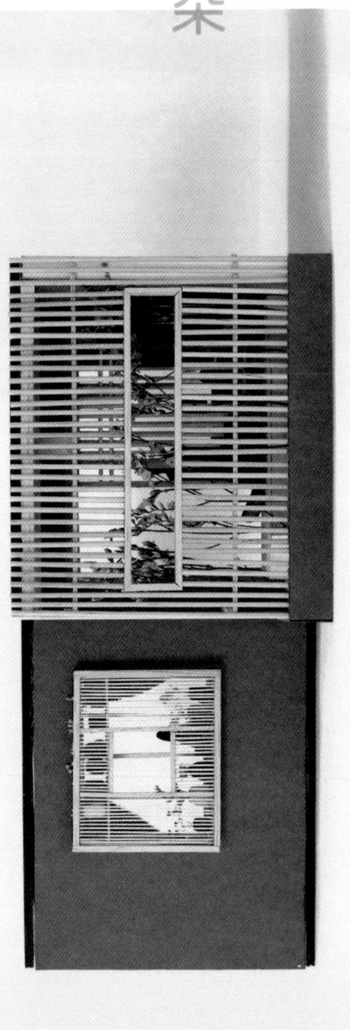

Apartment Blossom 楼房花朵 16

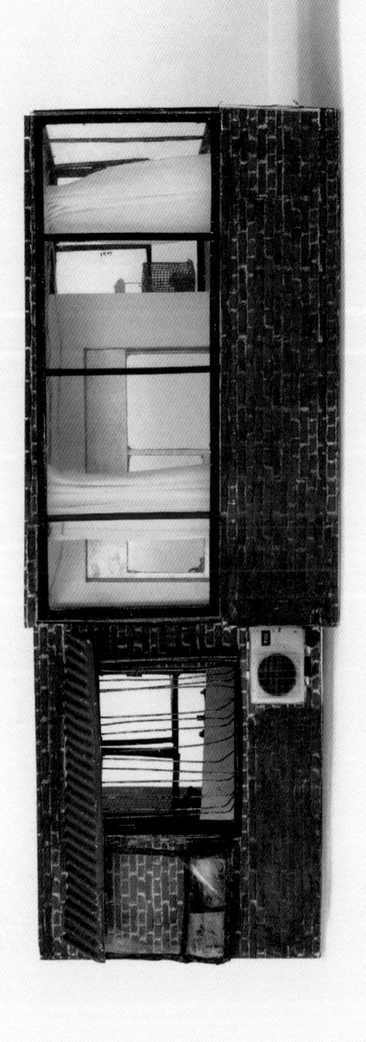

Apartment Blossom 楼房花朵 23

Apartment
Blossom
楼房花
朵
24

Apartment Blossom 楼房花朵 25

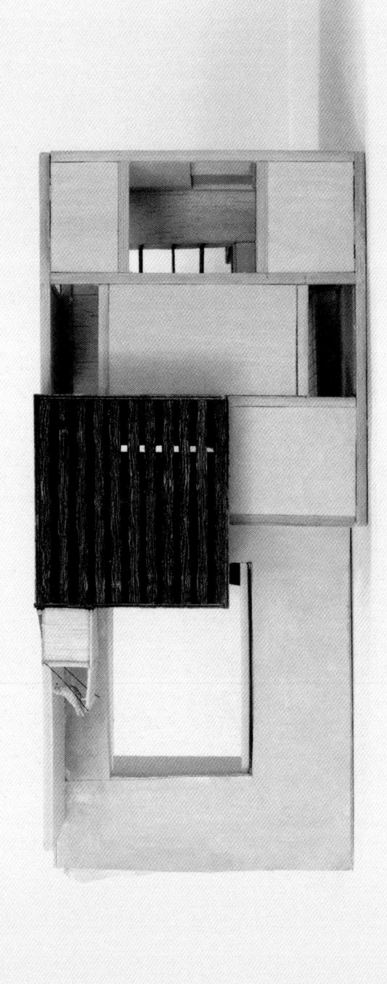

Apartment Blossom 楼房花朵 29

Apartment Blossom 楼房花朵 30

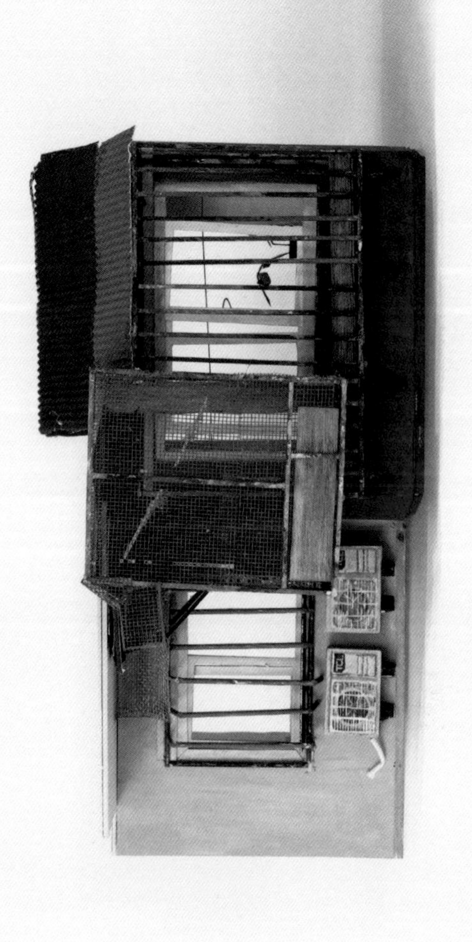

Apartment Blossom 楼房花朵 31

Apartment Blossom 楼房花朵 34

Apartment Blossom 楼房花朵 35

Apartment Blossom 楼房花朵 44

Apartment Blossom 楼房花朵 54

Apartment
Blossom
楼房花
朵
55

Apartment Blossom 楼房花朵 60

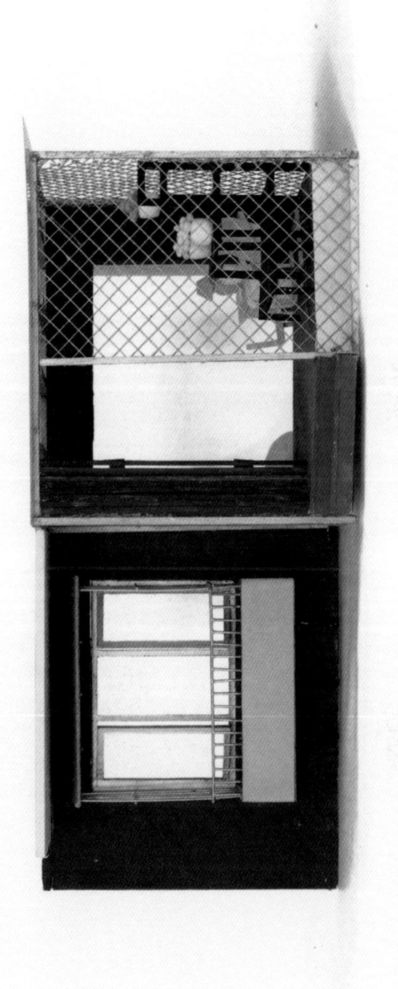

Apartment Blossom 楼房花朵 64

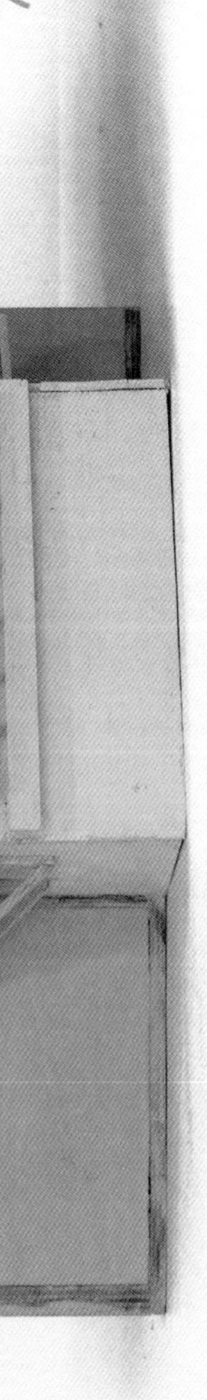

Apartment Blossom 楼房花朵 67

Apartment Blossom 楼房花朵 69

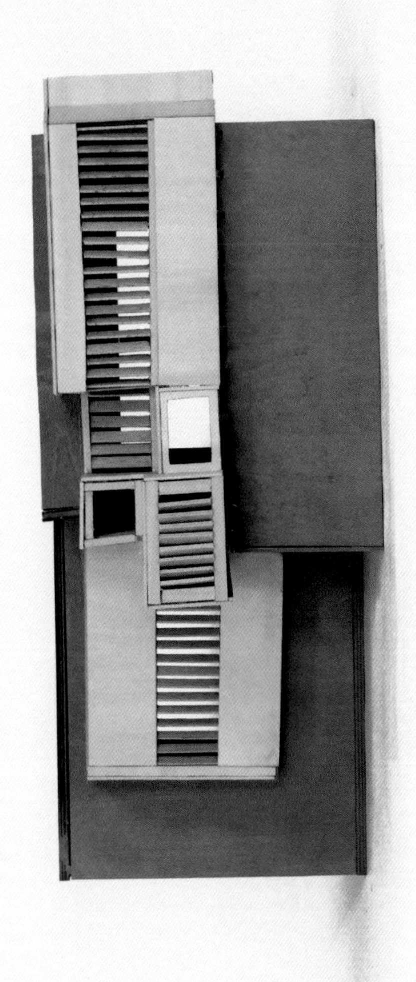

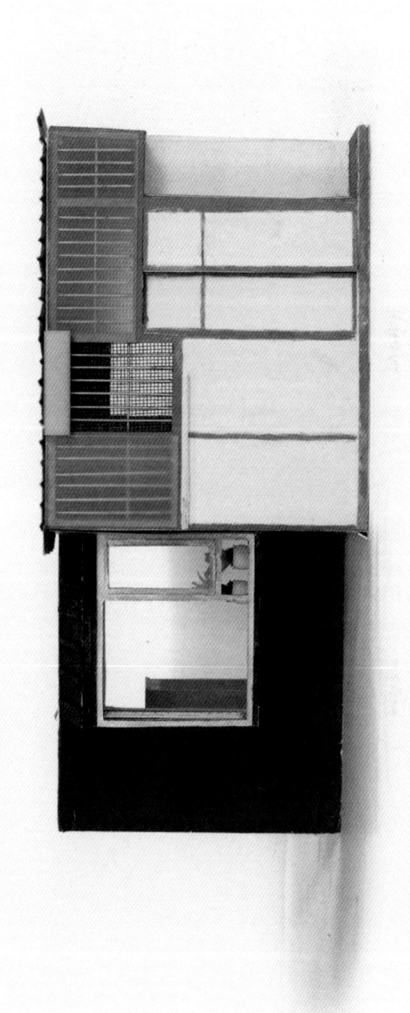

Apartment Blossom 楼房花朵 76

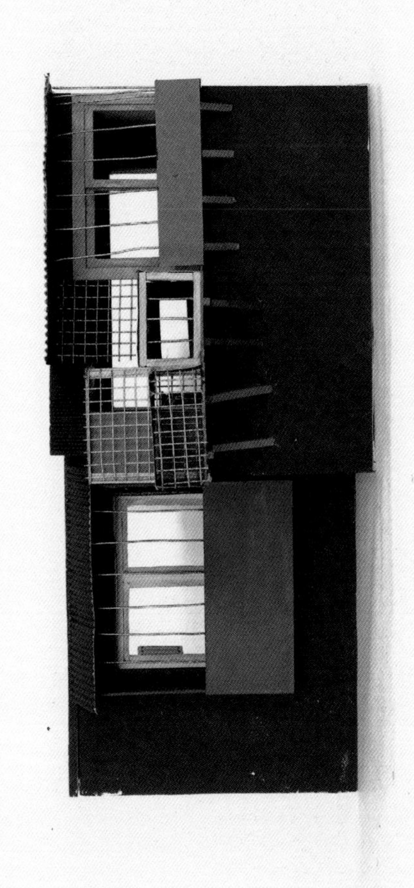

Apartment Blossom 楼房花朵 78

Apartment Blossom 楼房花朵 79

● Falling from the Balcony to the Miniature World

Li Han

Cliff Cages

I like taking walks in Beijing's old neighborhoods, which are visual feasts of folk construction. It abounds in the Hutongs as well, popping out from the corner of a yard, or up on a roof. No matter their dimension, most could still be considered houses. They are different in the old neighborhoods dominated by apartment buildings. There, they exist not to create inhabitable rooms, but to enclose windows and balconies. As a result, most such structures are simpler, more fragmented, and flatter, not much more than architectural elements. But their locations are quite conspicuous. When you look up, you can see them spreading all over the walls of apartment buildings. Some are protruding, some are recessed; sometimes they are connected in a row, sometimes they are scattered apart. The folk constructions in the Hutongs usually hide themselves away in corners, while the constructions here expose themselves with reckless abandon, like those pine trees on the Yellow Mountain, looking particularly joyful. With stricter governance in the last couple of years, many of the private constructions in the Hutongs were demolished, while those in the old residential areas were lucky to escape demolition and even continue to prosper. It is a little bit like the meteor hitting the Earth and causing the dinosaurs to go extinct, while the ants manage to survive thanks to their humble bodies.

Based on my observation over the years, I find the older the neighborhood, the more diversified and spectacular things there are to discover. The neighborhoods at Wang Jing and Xi Ba He (developed in the 1990s) are less interesting than those at Fang Zhuang and Liu Jia Yao (developed in the 1980s), while the latter are surpassed by those at He Ping Li and Jin Song (developed in the 1970s). But when compared to Bai Wan Zhuang (developed in the 1950s), the so-called "No.1 Residential Area of New China", all the other neighborhoods are no match. For some years, I worked at an office near the neighborhood, and would always spend part of my lunch break strolling through its streets, giving me the opportunity to observe how each household sealed its window and balcony. Some of them are aggressive, others are shy; some are clever, others are greedy. The total area of Bai Wan Zhuang is districted and numbered uniquely by the Twelve Earthly Branches (the most important counting system in the Chinese calendar) as District Zi, Chou, Yin, Mao··· As I remembered, some most splendid balconies and windows are from District Chou (a character meaning "ugly" in Chinese), which is well-deserved.

In the districts of Bai Wan Zhuang, you are presented with a history of window and balcony enclosures for residential buildings in Beijing. The oldest combination is the unenclosed balcony made of the original precast cement slab with windows characterized by warped wooden frames and broken single-pane glass, from which it can be surmised that the room is unoccupied. The more recent version is the balcony enclosed by steel framed windows. The slim steel frames divide the glass into small pieces, lightweight and sophisticated, proportioned to match the surrounding architecture. But their sealing performance is rather poor, so sometimes there will be another layer of aluminum alloy or plastic steel windows inside the steel framed ones. The owner of the house is probably some retired comrade who has been frugal throughout his life and feels guilty disposing of anything. So, in the end, the windows become double-paned! The bare aluminum alloy and plastic steel windows are the symbols of window enclosure in the 1990s. At that time, enclosing the balcony didn't simply follow the original boundaries of the architecture, so all kinds of complementary products emerged: the burglar-proof window, the canopy, the pigeon cage, the galvanized steel roof, the drying rack, making free combinations with windows and balconies and forming all kinds of overhangs. Since then, to enclose the window and balcony is no longer a two-dimensional concept, but a spatial action. The window and balcony are no longer some concave space on the façade of the building, but overhangs, like the cages on the cliffs. The latest enclosure configuration is a bottom-suspension casement system featuring bridge-cutoff aluminum alloy frames and three-layer hollow ultra-white glass with Low-E coating and built-in electronic blinds. It belongs to my old colleagues, a couple who are both architects.

Yoshiharu Tsukamoto published a book called *WindowScape* in which he collected, across many trips, all kinds of interesting windows from around the globe. The neighborhood of Bai Wan Zhuang is not large. But if the interesting windows and balconies from this area could be documented in a certain way, it could make another exciting book of its own.

Secret of Research

When Professor Jin Qiuye invited me back to lead another urban studies project, I pointed out of his office window to the neighborhood of Bai Wan Zhuang across the street: “That is it.”

Our research plan was as follows:

First, we brought more than 200 researchers to the districts of Bai Wan Zhuang and asked each of them to find one balcony and one window to document with a handmade model at 1:20 scale. This was the primary purpose of the research: to make an archive of the “cliff cages” at Bai Wan Zhuang in the form of models.

Based on those models, the research moved on to the second stage, in which the researchers were asked to redesign those balconies and windows following their own ideas so that more than 200 imaginative new “cages” were derived. This was meant to fulfill the researchers’ desire for design, although I wondered whether they would be able to surpass the residents at Bai Wan Zhuang.

In fact, we had a secret third stage but I didn't tell the researchers, as it might sound too ridiculous and unreasonable. It was that we would like to explore the "Degree Zero" of models. The concept of "Degree Zero" was discussed in the book *Hutong Mushroom*, the first volume of Urban Studies Degree Zero Series.

It originated from Roland Barthes and his book *Writing Degree Zero* (*Le degré zéro de l'écriture*). Barthes "regarded literature as 'myth', the significance of which was not to present the world but to present itself as literature."
Wang Junyang, The Everyday: A "Degree Zero" Agenda for Architecture (Part I), *Architectural Journal*, 2016-10, P23

A "Degree Zero" model is a model that presents itself as a model.

Model (All the models discussed in this article refer to physical models)

The discourse around models in recent years has been concentrated on their functionality: "Models should be more actively involved in the development of design...make many draft models...they can be observed more dynamically from different perspectives...Although the change of viewpoint in a computer model is similar to that of a physical one, [images] are always flat without depth. By elaborating [designs] through models···[one can get] an integrated spatial experience. [So] using a physical model to study massing, layout and scale is usually more accurate...The physical properties of a model can also be used to roughly verify the effect of structure, material, and light, which can give you a very concrete pre-judgement...It is more time-consuming to make a model, but it also depends on how one understands 'efficiency'. Working faster doesn't necessarily mean being efficient. Quality matters too. If a model can help to judge some issues more accurately, to improve the quality of design, or to prevent wasting effort constantly revising

Falling from the Balcony to
从阳台坠落
83

drawings, it should also be considered an improvement in efficiency." (Architecture Workshop)

Traditionally, models are used to represent a design on specific occasions, like a project presentation, competition, exhibition, forum, or archive. Compared to its status as a "design tool", being a "representation" sounds rather shallow, but the most exquisite model is usually created when the design is finalized and it is time to showcase its charm. Nowadays, with the accelerated pace of work, the development of industrial specialization, and the maturity of modeling techniques like 3D printing, the architect has given up his jurisdiction over the making of representational models and outsources their construction.

A model as such is seldom discussed by the architect. It sounds absurd to the architect that a model should not be used for elaborating design concepts or portraying an object, but should simply exist in its own right as the model itself, or as a display or artefact. Imagine, a model is created and says, "Look! I am a model. I am not to be used for making design. And I am not a miniature of anybody else. I am who I am!" Such a model would probably be thrown to the floor and crushed underfoot by the angry architect.

One day in 1885, Cézanne sat in the forest of the Fontainebleau and fell into contemplation in front of the scenery. He had too many choices for how to fill out the canvas: to catch the changes of light like Monet, to paint a gust of wind like Corot, to study perspective like Dürer, to tell a story of history like Repin, or to express nostalgia like Millet. In the end, he realized: "Everything is only 'painting', only the materiality composed of colors and brushes. Such materiality exists merely to prove the potential of painting materials, and that potential has nothing to do with using the material and means to describe the content. What the content describes now is the material and means – they are both merged into a unity. The pleasure of painting becomes the content of painting." Those words are extracted from

Retrospective of Retrospectives by the Chinese artist Chen Danqing, originally used to describe Manet's paintings. But wouldn't they be more appropriate to be used for the works of Cézanne, Matisse, and Pollock? The development of modern painting is just a process in which the paintings continued to return to "Degree Zero". For Manet and Cézanne, it was probably still only half conscious. For Matisse, it had already become conscious. And to Pollock, it became the whole premise. It has become a matter of common sense in modern painting, while it might still feel a little absurd to transfer it to the understanding of architectural models. However, we would like to give it a try.

Follow Cézanne's line of thought: a plaster model is just a piece of plaster, which is made by casting. We try to accentuate its sense of solidity and emphasize the traces of formwork. A wooden model is a block of wood, which is made by cutting. We try to accentuate the juxtaposition of horizontal and vertical sections. A paper model is a sheet of paper, which is made by folding. We try to accentuate its thinness and fragility. All such pursuits have nothing to do with elaborating a design or representing an object. What they relate to is the beauty of the medium and the model itself, which is the means to bring the model back to itself.

We requested that the researchers use mixed materials to make their models: cardboard or PVC board for the main structure and any available material on the market for the surface or components. We hoped the models could reinstate material reality as much as possible: metal for metal railings, wood for wooden fences, and a small piece of fabric for the hanging quilt. Although we had some hidden agenda, to be honest, I didn't know what the path would be for a model made of mixed material to return to its "Degree Zero". All that we could do was asking the researchers to relax their brains, open their eyes, get busy with their hands, and expect "Degree Zero" to appear. If there were several "Degree Zero" models or some hint of "Degree Zero" among more than 400 models, we would be satisfied enough.

Implication of "Degree Zero" of the Model

Paint

Two weeks later, most researchers had started to paint their models. We noticed that there were two groups: one was careful and cautious, painting their models with fine brushes and a single thin coat of paint. The color surface looked very even without any hint of brush strokes. The other was loose and generous, using big brushes and thick paint to cover the models over and over. The surface of their models looked like a painter's palette mixed with many different colors and obvious brush strokes. At first, I wondered if the buildings chosen by the cautious group were newer and cleaner while the loose group's were older and more ragged. But the assumption was quickly overturned. Even if it was an interior wall that was clean and white, the second group would still use big brushes to paint multiple coats. Even with a single color, there would still be the thickness of the paint coats. Don't overlook this subtle thickness. It was the condensed layer of the paint. "If you can perceive the beauty of texture from the condensed layer of the oil paint, that means you understand oil painting." Chen Danqing, *Local Perspective*, Season 2, Episode 10 Such textures admittedly give the model a painterly kind of beauty. But more importantly, they indicate the scale of the model. If we want to leave the same brush stroke and thickness on a 10 m high wall, how big shall the brush be? Can the painter still handle it? How thick must the paint be? Can it still stick onto the wall? If it is applied on a 10 cm-piece of cardboard, everything would be much more manageable. So, using big brushes and thick paints is not just a style, but a way to make the model look like a model. In their earlier practice, Morphosis always squeezed paints onto the models and used cardboards to scratch it on in order to create special paint textures. I used to think those models were very cool and unique. Now I feel they are even more like models.

Distortion

One day, I was drawn to a crooked model and stared at it for quite a while. The window, door, balcony and roof were all twisted, as if they had just experienced an earthquake. The researcher looked at me, sweating nervously. I hastened to say: "It is so beautiful!" I was not comforting him. Accidentally, he had reached the "Degree Zero" of the model.

Those twisted components were not made on purpose, but resulted from stress. The cardboard covered with paint was distorted. The thin paper crumpled up because of the moisture. Small parts were stretched due to the change of tension after different glues dried out. An originally proper model suddenly looked quite furious.

In the microscopic world of models, stress is more like the hand of God which gives life to materials, helping them to breathe and change. Especially for those relatively fragile materials like paper, corrugated sheet, cardboard, thin wooden sticks, plastic sheet, cloth, and tape—they are the most sensitive to stress, so they present the most abundant change. The "Degree Zero" of a mixed material model is probably the beauty of distortion created by the stress each material exerts on the others once pasted together.

Gehry's models are filled with the beauty of stress. He uses fragile materials to make models. The curves in his architecture are not drawn by hand, generated in Maya, or written with code, but are created by crushing paper and letting it open up little by little. Of course, that story is mostly apocryphal. But what cannot be denied is that Gehry's curves are the curves that result from the stress of materials in the microscopic world. It can be said that Gehry's architecture has amplified the stress from the microscopic world to the macroscopic scale. Architecture becomes the model of the model.

Thick & Strong

I saw a researcher working on a handrail for quite a few days. It was a kind of stainless-steel cylindrical handrail that is cheap, ugly, and can be found anywhere in the market. I picked it up and took a close look. The craftsmanship was wonderful. He used some plastic pipes and heated the material to join the elements, which is exactly how the stainless-steel ones are welded in reality. He even made the caps for the ends.

"The only problem is that your handrail is twice as thick as the real object. Correct scale is the top priority, above all else. We have stressed this point over and over!"

He said he knew his handrail was too large but he chose to do so anyway. The diameter for the actual stainless-steel pipe is 4 cm, or 2 mm if converted to the scale of the model. But he couldn't find any 2 mm plastic pipes. The thinnest one was 3 mm. He also thought about using tooth sticks, but they cannot be "welded" and are also not as smooth as plastic pipes. Between scale and texture, he gave up on the former and chose the latter, along with craftsmanship. As the elements were enlarged, the total size of the handrail became proportionally bigger.

After hearing his reasons, I looked at the handrails in my hand and realized that I was wrong. What we make is the model, not any simulation of another object. Therefore, when it comes to a decision between two competing priorities, the starting point of the model of course takes precedence over that of the object. The resulting thickness and strength in scale are actually correct within the world of the model.

The fruit pit boat—a fruit pit whose natural contours are carved to resemble a tiny boat—is a traditional motif for miniature sculpture. In his story *A Peach Pit Boat*, Wei Xueyi from the Ming Dynasty described the fruit pit boat carved by the master craftsman Wang Shuyuan as the following:

"Altogether, there are five people, eight windows, a canopy made of bamboo leaves, a paddle, a stove, a pot, a scroll painting, a string of prayer beads; ...However, the boat itself is less than an inch long." One must be extremely delicate to carve so many objects into a peach pit with a length of 3 cm. Today the most famous fruit pit boat, a masterpiece by Chen Zuzhang from the Qing Dynasty, is in the collection of the National Palace Museum in Taipei. However, when I saw the high-resolution image of this piece on the website, what struck me was not its fineness and delicacy, but its thickness and simplicity. Whether it was the body of the hull and the roof of the boat or the window frames and pillars, the forms were all very full-bodied. Compared to the scale of the wooden boat in the real world, the fruit pit boat in the microscopic world looks quite bulky. I understood that it was because of the way we see. When looking at the actual fruit pit boat, I feel exactly the same as Wei Xueyi. We only see its delicacy when looking into the microscopic world from the macroscopic one. But the photo enlarged the pit boat, or rather reduced me, so that I was able to observe the microscopic world with microscopic eyes, and therefore able to discover that bulkiness is the right scale for the microscopic world. The bulkiness is likely the result of material limitations (only 3 mm plastic pipe is available). But more importantly, it is due to the hand. Nimble fingers become clumsy giants in the miniature world, revealing a unique rustic quality and candid bulk within the "exquisite" world of miniature models.

Moe

On a sunny afternoon, I stepped into a south-facing classroom and my eyes were dazzled by the glare of sunlight reflected off a model. When I came closer, I couldn't help blurting out: "Kawaii!" (Japanese for cute). I didn't know where the researcher was, so I sat down to take a closer look.

This was a model for Phase Two, in which the researchers were asked to enclose the balconies and windows based on their own design. This imaginative "cliff cage" expressed the joy of a bird rushing out of its cage to regain freedom. Architecture, window, balcony—just let them go. "I just like the store signage, noren, hiragana; the origami garland, cartoons, coffee and desserts; the beautiful looks, flying dresses, and gentle water." The word here is not "model". Instead, it is "dollhouse".

It is said that dollhouses have been popular in the European court for hundreds of years. Little princesses probably played with them to dream up the love nests they would share with their Prince Charmings. Today it has become a giant industry, and the third most popular hobby in the States. No one would think to actually build a dollhouse. It would just be silly! We make a dollhouse to create a miniature world that we can always carry with us and place in the safest corner of the room wherever we go. If we have a family, we can pass it down to our children. To this point, the dollhouse is definitely a "Degree Zero" model. People expect it to stay miniature forever, and to never grow up until it becomes a family heirloom.

My random thoughts were interrupted by a huge shadow in front of me. A heavyset boy stood there with a pack of snacks. "Did you make this?" I asked. He nodded and smiled shyly. Well, a dollhouse made by an otaku; a house inside a house.

Cyberpunk Bai Wan Zhuang

Four weeks later, more than 200 researchers put together more than 400 models they had made. Every six models were stacked vertically as a group and all of them extended along the four walls of the exhibition hall. But we apparently underestimated the output. The models poured out the entrance of the hall, continuing down the corridors and disappearing into the crowd.

This was a six-story apartment building evolved out of windows and balconies that would stretch half a kilometer at full scale. Half of the balconies and windows were reproductions of reality, while the other half were the designs of the researchers, in which personal imagination, master techniques, and folk wisdom were all intermixed.

The reproductions of reality sometimes seemed more fantastical. The inner imaginations could feel rote in comparison. The absurdity of their total combination, the realness and fakeness of the details, made me feel that for a moment I was in the District Chou of Bai Wan Zhuang and had fallen into some two-dimensional cyberpunk world. It somehow felt like looking at a portrait by Picasso. All the facial features are figurative, but misplaced. Staring at it for a while, you would believe that the portrait looks so much like the model. But after a while, it doesn't even look like a human. Picasso used to comfort his model: "Ma'am, it is okay if you don't find the portrait to be like you. You will become more and more like the portrait over time." I wonder, will Bai Wan Zhuang become more like the models over time?

Exhilarated researchers were everywhere in the exhibition hall and corridors. They whispered and pointed fingers at the models, almost causing a "building collapse" many times, or asked friends to pose for group photos with the models. Looking at how joyful they were, I suddenly realized that they didn't know our research had a secret Phase Three. But at this very moment, all the researchers had unintentionally entered the status of "Degree Zero": imagining themselves reduced to 1/20th of their actual size, falling down from balconies or windows into their models, and enjoying this familiar yet strange miniature world.

Extreme Maquettes

The discussion of the "Degree Zero" of models is not meant to create a new type of model, nor even to oppose models as tools for design and representation. In fact, many working models are also excellent representation models, and at the same time standard "Degree Zero" models. The discussion about "Degree Zero" is meant to add a new layer to the thinking on architectural models such that more in-depth exploration can be carried out within that miniature world. For example, a rockery could be regarded as an extension of the natural scenery of a garden. But taking "fakeness" itself into consideration, the artificial mountain becomes full of life when it consciously responds to the diminutive scale of the garden. Congolese artist Bodys Isek Kingelez's talent is creating fantastical models of architectural and urban scenes. None of his works have actually been built, but they have been acquired for the permanent collections of many museums in Europe and the US. Kingelez calls his works "extreme maquettes". As a French word, "maquette" means both sketch and model, which reflects the model's double identity as both tool and representation. "Extreme maquettes" are supposed to push the sketch and model to their extremes, to "Degree Zero", and to become independent and complete works in and of themselves. It is both a sketch and a finished work, both a process and an end, both representational and independent. The ambiguity and contradiction of "extreme maquette" makes the model more complex and fascinating. It gives the model a broader perspective and grander ambition.

Models for Healing

I'm guessing you don't know who Peter Fritz is. In the summer of 1993, an artist named Oliver Croy wandered into a vintage store in a small town in Austria and found two large boxes in the corner. He was stunned when he opened them. Inside, carefully stacked, were 387 hand-made cardboard models. He bought them all for a very small sum. Based on the store's records, the models were made by someone named Peter Fritz. Croy began searching for information on Fritz, but could find nothing more than that he had been a clerk at an insurance company. Maybe making cardboard models was his hobby, his way to kill time on weekends or at night after work. Nobody knows how those 387 models he made throughout his life ended up in the vintage store. In 2013, the Venice Biennale invited Croy to exhibit his collection of Fritz's models, and they evoked an enormous outpouring of emotion. I saw the photos of the exhibition online. The innocence showcased in those models imprinted on my mind. I always thought they had been the work of some architect until writing this article. I remembered those models and went back to read the introduction to them. Compared to the models, the story behind them is far more extraordinary. Fritz reminded me of Kafka, who also worked at an insurance company, and who never published any significant works before his death, but became a literary legend afterward. Fritz probably never built any house while he was alive, and he remained unknown after his death. Kafka expected his scripts to be burned, but his friends had them published. Fritz's models were sold as waste, but were accidentally discovered and exhibited at the Venice Biennale, as fate would have it. Every time I've felt confused and anxious, I love to take a look at Fritz's models. The innocence and happiness presented in the models is the best healing.

● 从阳台坠落到袖珍世界

李涵

峭壁箱笼

我喜欢在北京的老小区散步，那是一场民间建造的视觉盛宴。在胡同里也可以欣赏到很多民间建造，从院子的某个犄角旮旯儿或是屋顶上冒出来，大小不一，但多数还算个房子。老小区的民间建造不一样，它们的存在不是为了盖房子，而是为了封窗户和阳台，因此大都更简单、更片段、更扁平，充其量算个建筑元素。但是它们的位置非常显耀，抬头一望，居民楼的墙壁上到处都是。有的凸出来，有的凹进去；有时连成一排，有时铺成一片。胡同的民间建造都是猫着藏着的，而这里的就像黄山上的松树肆无忌惮地袒露着，看起来特别过瘾。这两年随着治理严格，胡同里的民间建造拆得七七八八了，而老小区的倒是幸免于难，甚至还在欣欣向荣地生长。这有点像陨石撞地球，恐龙灭亡了，但是蚂蚁以其卑微的身躯，反而能苟且偷生。

这些年逛下来，我发现越老的小区，可看的越丰富越精彩。望京、西坝河（20世纪90年代建成）没有方庄、刘家窑（80年代建成）好，方庄、刘家窑没有和平里、劲松（70年代建成）好，但这些小区要是跟“新中国第一住宅区”百万庄小区（50年代建成）比起来，就都小巫见大巫了。有几年我在那附近上班，每天午休就在百万庄小区里溜达，看各家封的窗户阳台：有的嚣张，有的扭捏，有的巧于因借，有的贪得无厌。百万庄是绝无仅有按照“子丑寅卯”来分区的，我记得最精彩的几个阳台和窗户都在丑区，可谓实至名归。

在百万庄小区，你可以看到一部北京居民封窗户阳台的历史。最老的组合是留着原有水泥预制栏板，压根儿没有封起来的阳台，加上变了形的木窗和破裂的单玻，估计已经没人住里面了。比较老的是钢窗封的阳台窗户，纤细的钢框划分出小块玻璃，轻盈典雅，与建筑的比例最协调。但是密封性不好，所以钢窗里边有时还有一层铝合金或者塑钢窗。估计是老同志住里边，节俭了一辈子，扔东西是罪孽，最后变成了双层窗！赤裸的铝合金窗与塑钢窗是90年代的封窗符号。这时候封窗户阳台已不是简单地卡着边界，一封了之了。配套产品已经出现：防盗窗、雨棚、鸽子笼、彩钢顶、晾衣架，与窗户和阳台自由组合，形成各种出挑。封窗户阳台从此不再是一个平面概念，而是一个空间动作。窗户阳台不再是建筑立面上的凹入空间，而是突出物体，犹如峭壁上的箱笼。最新的封窗配置是断桥铝三层中空超白玻璃内置电动升降百叶外镀Low-E膜下悬平开系统窗，主人是我的前同事：一对建筑师夫妇。

塚本由晴写过一本书叫《世界之窗——窗边行为学》（*WindowScape*），里边是他踏遍千山万水，环绕地球搜集来的各种有趣的窗户。百万庄不大，但是如果能够把里边各种有趣的窗户阳台以某种形式记录下来，也一定能凑出一本精彩的书。

研究的秘密

当金秋野老师又一次邀请我带一个城市研究课题时，我毫不犹豫地指着他办公室窗外马路对面的百万庄小区："就它吧！"

我们的研究计划是：

首先把200多位研究员撒到百万庄子丑寅卯各区中，每人找到一个阳台一个窗户，用1:20的手工模型记录下来。这是研究的基本目的：用模型为百万庄的"峭壁箱笼"留下一份档案。

在此基础上，研究进入第二阶段，研究员们可以按照自己的构想把这些阳台窗户重新封一遍，衍生出200多个想象的"箱笼"。这是为了满足研究员的设计欲望。我担心他们会输给百万庄小区的居民。

我们其实还有一个秘密的第三阶段，但鉴于它听上去过于荒谬，不可理喻，我们没有向研究员公布，那就是我们想研究一下模型的"零度"。"零度"这个概念在《零度城市研究系列》的第一本书《胡同蘑菇》中进行过讨论："零度"概念源自罗兰·巴特（Roland Barthes）和他的《写作的零度》。他"将文学作为一种'迷思'（myth），其意义首先不在于呈现世界，而是将自己呈现为文学"。王骏阳，日常：建筑学的一个"零度"议题（上），《建筑学报》，2016-10

"零度"模型就是将自己呈现为模型的模型。

模型（本文里的模型均指实体物理模型）

近年来对模型的讨论集中在工具性方面："模型要更多地参与到

设计的推进中去……做很多草模……模型可以从不同视角更加动态地去看……电脑建模中的视点转换尽管跟实体模型接近，但（画面）总是扁的、无深度。而在模型上推敲时……是很整体化的空间感受。（因此）用立体的模型去推敲体块、空间的布局和比例往往来得更准确……模型本身的物理属性也可以去粗略地验证结构、材料、光的效果，它能够给你一个非常实在的预判。……模型做起来要更费时，但是也要看怎么理解‘效率’，不是说工作进度越快就越有效率，质量也很关键。模型如果能更准确地去判断一些问题，提升设计质量，省去不断修改图纸的无用功，那也是一种效率的提升。”（建筑工房）

传统上，模型更多地是用来表现设计的。项目汇报、竞赛投标、展览论坛、档案记录都少不了表现模型的身影。尽管比起“作为设计的工具”，“表现”是个相当肤浅的事儿，但是最精美的模型往往都是设计尘埃落定，需要充分展现方案魅力时诞生的。如今随着工作节奏加快，行业分工细化以及3D打印等模型制作技术的成熟，表现模型在越来越大的程度上离开建筑师的管辖，沦为外包工作。

作为模型的模型很少被建筑师讨论。模型既不用来推敲设计，也不用来表现对象，而是作为模型本身而存在，或者说作为一个摆件、一个器物而存在，这对建筑师来说是荒谬的。一个模型被制作出来就是为了说：“看！我是个模型。我不是用来做设计的，我也不是谁的缩小物，我就是我自己！”这样的模型大概会被愤怒的建筑师掀翻在地，再用脚踩个粉碎。

1885年的一天，塞尚（Paul Cézanne）坐在枫丹白露的森林里面，对眼前的风景陷入深深的思考。他有太多的选择来填满画布，像莫奈（Claude Monet）那样捕捉光线的变化，像柯罗（Jean Baptiste Camille Corot）那样画一阵风，像丢勒（Albrecht Dürer）那样研究透视，像列宾（Ilya Yafimovich Repin）那样讲述一段历史，或者像米勒（Jean-Francois Millet）那样抒发一段浓浓的乡愁。最终他想明白了：“一切只是‘画’，只是色与笔构成的物质感，这物质感仅仅为了证明绘画材料的可能性，而这可能性不再是材料手段诉说内容。内容，现在诉说的是材料和手段——二者合一，绘画的快感乃成为绘画的内容。”这段文字摘自陈丹青的《回顾展的回顾》，原文用来描述马奈（Édouard Manet）的绘画。但放在塞尚、马蒂斯（Henri Matisse）、波洛克（Jackson Pollock）的绘画上不更合适吗？

现代绘画艺术的发展正是一个不断回归绘画“零度”的过程，在马奈、塞尚那里还是半自觉，到了马蒂斯就是主动自觉，到了波洛克就是基本前提了。这道理在现代绘画上早已是常识，移植到建筑模型上竟然有点荒谬，可我们想试试。

按照塞尚的思路：石膏模型就是一块石膏，它是被浇筑出来的，玩儿的就是整体性与拆模的痕迹感；木模型就是一块木头，它是被锯出来，玩儿的就是木材横切面和纵切面的变化组合；纸模型就是一张纸，它是被折叠出来的，玩儿的就是纸的薄和脆弱感。这些追求与推敲设计无关，与表现对象无关，它们关乎的是媒材的美，模型自身的美，是让模型回到模型的途径。

我们要求研究员制作的模型是综合材料：主体结构是卡纸板或PVC板，表皮或构件可以是市面上能淘到的任何材料。我们希望模型要尽可能地还原真实情况。金属栏杆就用金属做，木栅栏就用木头做，晾的被子就找一小块布头。虽然我们心怀不可告人的“秘密”，但说实话，一个综合材料的模型回归“零度”的途径是什么，我自己也不清楚。我们能做的就是让研究员把大脑放松下来，眼睛敏感起来，双手勤快起来，等待“零度”出现。如果能够在400多个模型中发现几个零度模型或是几处“零度”的意味，我们就心满意足了。

模型的零度意味

涂抹

两周后，大多数研究员进入涂色阶段，我们注意到涂色分成两派：一派小心谨慎，细笔薄涂，一遍完成。模型颜色均匀，看不出笔触。另一派粗放豪爽，大笔厚涂，反复覆盖。模型表面像个调色板，混合着各种颜色，有明显的笔刷痕迹。我一开始疑心是不是因为谨慎派选择的楼比较新、比较干净，粗放派选择的楼比较旧、比较破烂。但这个假设很快被推翻。粗放派即使刷一堵室内干净的白墙也会大笔来回涂抹，即便没有颜色的混合，也会留下颜色的厚度。不要小看这层微妙的厚度，这可是颜料的凝结层。“你能看到油画颜料凝结层的质地美，你就算看懂油画了。”陈丹青，《局部》，第二季第10集 这种质地固然让模型具备了某种绘画意义上的美，但更为重要的是，它们提示了模型的尺度。如果在一堵10米高的墙上留下同样的笔触和厚度，需要多大的刷子？油工还能刷得动吗？需要多厚的涂料，它还能挂在墙上吗？但在10厘米高的纸板上，一切挥洒从容。所以大笔厚涂不只是一个风格，而是使模型看上去就是模型的方法。墨菲西斯（Morphosis）早年做模型时，把颜料挤到模型上，再用卡纸板刮抹，产生特殊的颜料肌理。当年我觉得那些模型很酷很特别，现在觉得它们更像模型了。

变形

有一天，我被一个歪歪扭扭的模型吸引了，凝视了许久。窗户、门、阳台、屋顶都是歪的，好像刚经历过一场地震。研究员紧张地看着我，额头似乎在冒汗。我赶忙说：“太美了！”我不是在安慰他，他无意中做出了模型的“零度”。

这些倾斜扭曲的构件不是因为他要做成那样，而是应力带来的变形。纸板附着了颜料发生的扭曲，薄纸片因潮湿发生的翘角，还有各种胶在凝固后产生的拉力变化对于小零件的撕扯，让原本一个规矩方正的形体突然张牙舞爪起来。

在微观的模型世界里，应力就像上帝之手，赋予材料生命，让它们“呼吸”、变化。特别是那些相对脆弱的材料，纸、瓦楞纸、卡纸板、细木棍、塑料片、布、胶带，它们对于应力最敏感，变化最丰富。一个综合材料模型的“零度”，也许就是各种材料粘贴后在应力作用下产生的变形之美。

盖里（Frank Gehry）的模型充斥着应力美。他用脆弱的材料做模型。他的建筑曲线不是手画出来的，不是玛雅（Maya）拉出来的，也不是代码写出来的，而是把纸一团，纸再一点点张开出来的。这当然是个玩笑说法，但不可否认的是，盖里的曲线是一种微观世界的材料应力曲线。盖里的建筑，可以说是把微观世界的张力放大到了宏观尺度，建筑成了模型的模型。

粗壮

我看到一位研究员好几天都在做一个栏杆，而且是那种市场上到处都是，便宜又难看的不锈钢圆管栏杆。我拿起来端详了一会，做工真的精细。用塑料管做的，杆件之间是加热后让塑料融化粘到一起的，和现实中焊不锈钢的思路一模一样。连端头的扣盖都做了。

“唯一的问题是栏杆的比例比实物粗大了一圈。尺度正确优先于其他任何因素，这可是我们反复强调的重点！”

他说他知道做大了一圈，但他选择做大。真实的不锈钢圆管直径是4厘米，折合成模型就是2毫米，但他买不到直径2毫米的塑料管，只有3毫米的。他曾想用牙签，但没法“焊接”，也没有塑料管光滑。在比例和质感的选择上，他放弃了比例，选择了质感和工艺。由于杆件放大了，所以栏杆整体尺寸也适度拉大了。

听完他的理由，再看看手中的栏杆，我意识到我错了。我们做的是模型不是对象的模拟物，所以当遇到矛盾需要取舍时，模型的出发点当然优先于对象的出发点，而由此产生的粗大比例恰恰是模型世界的正确尺度。

核舟是微雕的传统主题，就是把果核雕刻成一叶扁舟。明代魏学洢作《核舟记》描述奇人王叔远雕刻的核舟：“通计一舟，为人五；为窗八；为箬篷，为楫，为炉，为壶，为手卷，为念珠各一；……而计其长曾不盈寸”。3厘米长的桃核上雕刻出这许多内容，想必是极度的细密精巧！今天存世的最著名的核舟在台北故宫博物院，是清代大师陈祖章的作品。然而当我在网站上看到这件作品的高分辨率照片时，我感受到的不是细密精巧，而是粗壮朴拙。无论是船身船篷还是窗棂柱子，形状都是丰腴饱满的，比起真实世界木船的比例，微雕世界的核舟显得五大三粗。我明白这跟观看方式有关，如果看的是实物核舟，我跟魏学洢的感受定是相同的，从宏观世界俯视微观世界，看到的必然是精细。但照片把核舟放大了，也可以说把我缩小了，我得以用微观之眼看微观世界，于是发现粗壮才是微观世界的比例。粗壮或许是因为材料限制（只有3毫米粗的塑料管），但更重要的原因应当是：手。灵巧的双手在模型的世界里成了庞然大物，有时笨拙得无从下手。这让“精细”的微观模型世界反而呈现出一种独特的拙朴感，一种憨厚的粗壮比例。

萌

某个晴朗的午后，我走进一间朝南的研究室，进屋就被一个模型反射的刺眼阳光晃着了。走近一看，禁不住说了句日语：卡哇伊！研究员不知道去哪儿了，我便坐下来仔细端瞧。

这是第二阶段模型，研究员按照自己的设想封阳台和窗户。这组想象的“峭壁箱笼”透出了一种鸟儿冲出笼子重获自由的快感。什么建筑、窗户、阳台都散了吧。本小姐喜欢的是店招、暖帘、平假名；是折纸拉花、卡通漫画、咖啡甜点；是容颜娇艳、衣裙漫飞、温柔如水。这里没有模型这个词，取而代之的是“娃娃屋”。

据说娃娃屋几百年前就开始在欧洲宫廷流行了，可能是小公主们用来构想与未来白马王子共筑爱巢的玩具。今天它是一个巨大的产业，在美国是排名第三的公众爱好。没人会真的想把娃娃屋盖出来，那样可太傻了！做娃娃屋就是为了打造一个迷你袖珍世界，并把它一直带在身边，无论走到哪里都放在房间最安全的角落。等到将来结婚生子，再传给下一代。从这个角度讲，娃娃屋是绝对的零度模型，人们就是希望它永远那么袖珍，永远不长大，直到成为传家宝。

我的浮想联翩被眼前一个巨大身影打断。一个白胖男生拎着一包零食站在面前。我问：“这是你的模型？”他腼腆地笑着点头。宅男做的娃娃屋：宅中宅！

赛博朋克百万庄

四周后，200多位研究员把他们做的400多个模型码放在了一起。每六个模型上下落成一组，沿着展厅的4面墙徐徐展开。但我们明显低估了工作量，模型冲破了展厅的入口，沿着走廊延伸出去，消失在人群里。

这是一个由窗户和阳台演化出来，按比例计算会延绵近半公里的6层居民楼。有一半的阳台和窗户是现实的翻版，另一半是研究员混合了个人想象、大师手法和民间智慧的设计。

现实的翻版有时看上去更荒诞，内心的想象有时倒比现实更苟且。整体组合的荒诞与亦真亦假的局部让我一会儿觉得身处百万庄丑区，一会又陷入某个赛博朋克的二次元世界。这有点像看毕加索的肖像画，眉眼口鼻全都是写实的，但位置全是乱的。盯着看一会儿觉得像极了模特本人，可是目光一收回来，连个人都不是。毕加索曾安慰他的模特说：“夫人，画得不像没关系，慢慢地您就会像这张画的！”我想，百万庄会慢慢地像这个模型吗？

展厅里、走廊外，到处都是兴奋的研究员，他们对着模型窃窃私语，指指点点，好几次发生"房倒楼塌"的险情，或者呼朋唤友摆出各种姿势与模型合影。看到他们手舞足蹈的样子，我突然意识到他们还不知道我们的研究还有个秘密的第三阶段，但是此时此刻，所有研究员已经不由自主地进入到了零度状态：想象着缩小了20倍的自己，从阳台或窗户坠落到模型中，尽情享受这个熟悉又陌生的袖珍世界。

终极的模型 / 草稿

对于模型的"零度"的讨论，并不是要划出一类新的模型，更不是对作为设计工具和表现工具的模型的反对。事实上，很多工作模型同时也是精彩的表现模型，同时还是标准的零度模型。对零度的讨论是想给建筑模型的思考叠加一个新的层次，从而可以在微观的世界中进行更深入的探寻。就像假山是自然山水在庭院中的投射，但对于"假"本身的思考，对于咫尺庭院的自觉，让假山变得更加生机盎然。刚果艺术家波迪斯·埃塞克·金格勒兹（Bodys Isek Kingelez）的绝技就是制作奇幻的建筑和城市模型。这些模型无一被实际建造出来，却成为欧美各大美术馆的收藏。金格勒兹把自己的作品统称为"Extreme maquettes"（终极的模型）。"Maquette"是法语，既指草稿也指模型，双关了模型做为工具和表现的双重身份。"Extreme maquettes"就是要把草稿和模型推演到极致，走向"零度"，成为独立的、完成的作品。它既是草稿，也是完成品；既是过程，也是终点；既是再现的，也是独立的。"Extreme maquette"的暧昧与矛盾让模型更加复杂、迷人，它为模型提供了更广阔的视角和更大的野心。

作为疗愈的模型

你一定不知道彼得·弗里茨（Peter Fritz）是谁。1993年的夏天，一位叫奥利弗·克罗伊（Oliver Croy）的艺术家在奥地利小镇闲逛，他走进一家旧物商店，在一个角落里发现两个大箱子。打开箱子的瞬间克罗伊被震住了，箱子里整整齐齐码放着387个手工制作的卡纸模型。他以很便宜的价格买了下来。根据店家的资料，这两箱子模型是一个叫彼得·弗里茨的人做的。克罗伊开始查询关于弗里茨的信息，然而除了知道他是一名保险公司的职员，再无更多资料。做卡纸模型或许就是他的业余爱好，用来打发无聊周末和漫漫长夜。但他穷其一生攒下的387个模型如何卖给了旧物店，无人知晓。2013年，威尼斯艺术双年展邀请克罗伊将他收藏的387个弗里茨做的模型展出，引来无数震惊与感动。我是在网上看到这个展览的照片，那些模型展现出的天真烂漫让我过目不忘。我一直以为这是某个建筑师的作品，直到写作本文的时候，那些模型又浮现在眼前，才又翻回去细读了介绍。比起模型，这背后的故事更令人唏嘘。同为保险公司职员，弗里茨让我想起了卡夫卡（Franz Kafka）。卡夫卡生前未出版一部重要作品，身后成为一代宗师；弗里茨生前估计没有盖过一栋房子，身后依然默默无闻。卡夫卡希望自己的手稿被烧掉，但朋友把它们出版了；弗里茨的模型被当成废品卖掉，却被人意外发现，最终搬到了威尼斯双年展。命运弄人！每当困惑迷茫、焦虑烦闷之时，看一眼弗里茨的模型，那些天真美好，那些单纯快乐，真是最好的疗愈！

Falling from the Balcony to the Miniature World

从阳台坠落到袖珍世界

103

Falling from the Balcony to the Miniature

从阳台坠落到袖珍世

103

Apartment
Blossom
楼房花
朵
104
Apartment
Blossom
楼房花
朵
104

Apartment Blossom 楼房花朵 105

Apartment Blossom 楼房花朵 105

Apartment
Blossom
楼房花朵
106

Apartment
Blossom
楼房花
朵
107

Apartment
Blossom
楼房花
朵
108

Apartment
Blossom
楼房花
朵
109
Apartment
Blossom
楼房花
朵
109

Blossom
朵
Apartment
楼房花
110
Blossom
朵

Blossom
朵
Apartment
Blossom
楼房花
朵
111

Apartment
Blossom
楼房花
朵
112
Apartment
Blossom
楼房花
朵
112

Apartment
Blossom
楼房花
朵
114

Apartment
Blossom
楼房花
朵
115

Apartment Blossom 楼房花朵 116

Apartment Blossom
楼房花朵
117
Apartment Blossom
楼房花朵
117

Apartment
Blossom
楼房花
朵
119

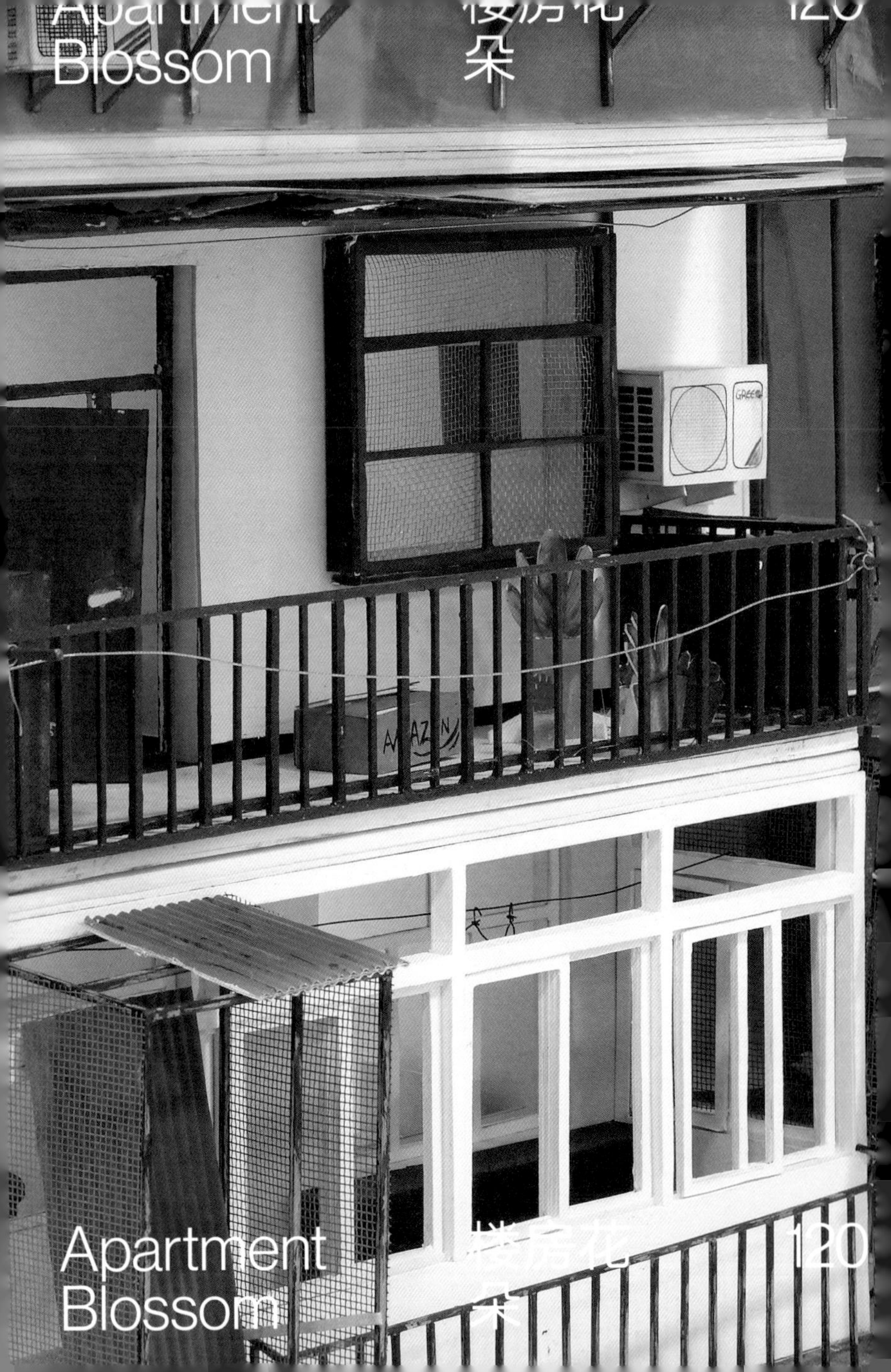

Apartment
Blossom
楼房花
朵
120

Apartment
Blossom
楼房花
朵
121

Mary Jane
Mary Jane
Apartment
Blossom
楼房花
朵
122

Blossom
朵
Apartment
Blossom
楼房花
朵
123

● Balcony, a stage of life.

Jin Qiuye

Alfred Hitchcock reproduced the inner courtyard of a typical New York City block in his masterpiece *Rear Window*. Neighbors peer across the courtyard into each other's windows—real life has always been more alluring and absorbing than fiction. But, after all, what can be seen through the windows is limited, especially with curtains. The balconies, however, would expose the private life more directly. In a sense, a balcony is a mirror of one's private life. No matter what clothes a person wears, their faces are always left uncovered, shaped into social expressions. It's the balcony that wears a home's expression. Through the lens of *My Brilliant Friend (L'amica geniale)*, we studied poor Neapolitan neighborhoods in the 1950s. Balconies overlooking inner courts were not just extensions of private space, but also places where neighbors, particularly housewives, communicated across the open air. When one woman, in the heat of an argument, chucked item after item from her balcony, private life was exposed to public space. The inner court reverberated with her clear and powerful wails as others on their own balconies stared speechless at each other. The balcony turned into a stage upon which the comedies and tragedies of life were all pushed into the spotlight.

In contemporary apartments, balconies have replaced windows as the links between private space and public life. The urban residence with a balcony is a particular residential type, redefining the relationship between inside and outside within a city. For instance, Beijing's traditional quadrangle, or *siheyuan*, could be understood as a courtyard-style villa. In a villa, enclosing walls and locked gates clearly mark the boundary between inside and out. Many high-end villas are designed on the premise of this hard boundary. There is no public life within the residential walls. But a *dazayuan*, a *siheyuan* occupied by several different households, is quite the opposite. The courtyard space is frequently filled with parents and kids making their rounds. Their private life is no longer enclosed. Instead of space, time becomes the means of separating private and public life: residents can only enjoy peace and privacy when they shut their door at night. In contrast, high-density urban multifamily housing makes a good balance of these two kinds of courtyards. It enables public life to infiltrate the overall space, while letting private life stay hidden behind burglar-proof doors, only showing its true colors on the balcony.

In cold northern regions, balconies are always enclosed. A completely enclosed balcony is different from an exterior window. There is no thermal insulation layer, and the opening is larger in comparison. As an indoor space, the balcony is still seen as a transitional space, a link between inside and outside. The following is Walter Benjamin's childhood recollection of the enclosed balcony in his grandmother's home:

"The most important of these secluded rooms was for me the loggia. This may have been because it was more modestly furnished and hence less appreciated by the adults, or because muted street noise would carry up there, or because it offered me a view of unknown courtyards with porters, children, and organ grinders. At any rate, it was voices more than forms that one noticed from the loggia. The district, moreover, was genteel and the activity in its courtyards was never very agitated; something of the insouciance of the rich, for whom the work here was done, had been communicated to this work itself, and a flavor of Sunday ran through the entire week. For that reason, Sunday was the day of the loggia. Sunday—which the other rooms, as though worn out, could never quite retain, for it seeped right through them—Sunday was contained by the loggia alone, which looked out onto the courtyard, with its rails for hanging carpets, and out onto the other loggias; and no vibration of the burden of bells, with which the Church of the Twelve Apostles and St. Matthew's would load it, ever slipped off, but all remained stored up in it till evening."

For adults, the balcony from inside is a window to the outside world. For children, the balcony is where they independently face the world. A balcony makes the outside world accessible, but also cuts it off. Even if it does not visually dominate the home, as something that belongs to both the exterior and the interior, it brings inside with it just the tiniest sliver of nature. In this way, a balcony seems not unlike an open patio or inner courtyard, serving as a protective buffer between living spaces and the everyday life outside.

As both an indoor and highly functional unit, the enclosed balcony rivals the appeal of terraces, porches, roof gardens, even in Buddhist monastery courtyards and hermits' mountain and forest retreats. It represents a step beyond the bread and butter of life, carrying with it an aspirational quality. It is one of the great contradictions of home life. To say "I'd like a large balcony" means you are admiring something that you can never actually acquire.

There was a story in early July: a sleeping tenant fell from a tenth-floor balcony because the landlord had converted the balcony into a bedroom. Such balcony collapses have happened on more than one occasion, in one case after a landlord converted the balcony into a toilet. These are probably the most extreme cases of reconstruction. However, balconies still perform an enormous variety of semi-outdoor functions, serving as a playground, gym, greenhouse, vegetable patch, dog run, aquarium, laundry room, clothesline, and storage shed. All of the above functions that are usually integrated into the courtyards must be adapted to this transitional space barely one meter wide. With such a variety of functions and purposes, balconies as a type grow rich in characteristics and vivid details.

Another use for the balcony is demonstrated in my project Doctor Li's House. The balcony was designed to be a fully indoor space, a tea room or living room. With indoor space thus expanded, new problems followed: the balcony had

Balcony 透明的 126
A Stage of Life 舞台

been transformed into a core space in the house, ceasing to carry merely auxiliary functions. Such functions had to be transferred to other spaces: clothes drying in the bedroom or study, bicycles stored in the hallway, etc. This wouldn't be much of a problem in a big house, but for a small house, it could be an issue. The biggest disadvantage for a small house is the lack of complexity, making it feel like a small rock in a vast sea. It seems impossible to feel depth, inside and outside, in a narrow space.

In short, if the house is the quintessential form of architecture (from time immemorial, the core impulse for architecture has always been shelter), the significance of the balcony in architecture is that it sticks to the house. It is the family's public showcase, as well as a transformable multifunctional space. Since the balcony is an extension of the interior living space out toward the exterior, it has been an ideal way to carve out additional usable space. The building elevation is rich in forms ranging from extended pigeon coops to flower boxes and bay windows. Chinese people are used to simply expanding their homes outward, so they pay no mind to their wasteful and inefficient uses of indoor space. Balconies in an average city apartment are no more than 5 m^2, room for only the tiniest of spatial operations. They become the inevitable destination of household equipment like ventilation systems, dryers and electric clothes drying racks. Cleaning products, hardware, old electric fans, desk lamps, walking sticks, roller skates (last used years ago), spare floorboards and wall paint, house plants, and empty litter boxes all eventually collect there, too. Architects should use all the skills at their disposal to provide means to neatly organize or conceal this accumulation. If not, the versatile balcony space becomes nothing more than a necessary compromise to sustain home life.

In order to avoid the visual clutter of such compromise, the U.S. enacted strict laws and regulations to limit the types of items allowed on the balcony. Additional compulsory measures were adopted to ensure a neat and uniform appearance: balcony rails were required to be covered by black netting. Moreover, residents were not allowed to place anything on balconies that are connected to fire escapes. Even smoking on the balcony is banned—there's a designated smoking area in apartment buildings. People voluntarily give up singing and playing the guitar on the balcony out of respect for their neighbors. Once the balcony's functions are restricted in these ways, its usage is greatly reduced. People rarely emerge, and this absence of life turns the balcony into a fake *life landscape*.

So what is a real *life landscape*?

Many Asian cities still retain a sort of *ecological community*. Deeply rooted traces of life from over the years fill residential areas with random artifacts and layers of history. They are not all necessarily beautiful or pleasant. Normal urban life has both a polished front face and a dirtier flip side, just as there are always rotten plants, silt and dead fish beneath the surface of a pond. The ecosystem of a city can't be complete if its flip side is suppressed and only the polished face is left to surface. In the United States, not only is there little opportunity for the flip side to show its face in most suburban areas, but people try even to actively eradicate it. As a result, manicured lawns are seen in front of each and every household. Instead of symbols of *life*, these lawns become *landscape*. The link between inside and outside is cut off. There is little street life in the middle class residential areas in the United States. Runners or dog walkers pass in a hurry. There is no trace of children or the elderly. Most of the day, the street is shrouded in dead silence. This world is idyllic, but fictitious, and for me that's not real life. A *life landscape* should tell the full story of life, in all its aspects; it should embrace inclusivity and the complementary nature of both sides of life—they are inseparable. Where there are fresh buds sprouting, there are dead twigs and withering leaves. The superseding of the old by the new is the law of life, and there is no need to be ashamed of it.

The word "landscape" is getting stranger. Perhaps its meaning is developing with the world's ongoing digitization. When cities, infrastructure, the Earth, or anything vast in extent enters landscape design thinking, landscape is transformed from an object that can be physically perceived into a *concept*. Which is to say, scale is the key to landscape. For me, a concept is that which can't be physically perceived, nor can it be directly perceived through the senses. It's slippery, grasped only by heart and mind. Is this idea too personal? I feel safe assuming that all 6 billion people on Earth are mostly alike in their physical structure and sensory apparatuses. Therefore, personal feelings could be considered universal. This is not the prevailing attitude of the times. The underlying assumption behind Big *Data* is that personal feeling is neither real nor reliable. The future design of living environments will depend on perception at larger scales and more precise quantitative cognition.

I'd like to express my instinctive doubt of *concept*. The visible and tangible real landscape is the deck chair under the tree, the large rock on the mountaintop commanding a panoramic view, the scattered flower pots in the alley. It seems to me that the more distant the view, the more we feel and experience with our eyes in place of our bodies, the more picture it is than place. What is a life landscape? In short, innumerable narrow Hutong yards, public gardens, the balconies of every household shrunk and spread infinitely across a drawing. There is no magnificent topography or arresting terrain. They are closely packed and randomly distributed, each one with its own life stories. Order and chaos, the polished and tarnished, successfully coexist side by side. Such is a *life landscape*.

Life is not a show. It is usually hidden from public view. But when any part of life is materially exposed—a house with a balcony, a yard or a sidewalk garden, rather than just wide roads lined with trees and overpasses—it is staged for the public and takes on a dramatic effect. This visual presentation of life looks as beautiful as a green pond. The sea and the mountains are beautiful because of the ecosystems of life forms they support. Otherwise, there would be no value within the geography. Beauty cannot be created through concepts. The more individuals there are, the higher the diversity. These individuals together form an information rich spatial environment called a city. Is a city a form of big data? Without a doubt, it is the most significant form of big data we have today. Unlike traditional big data, the city doesn't require hard disk storage. Like a writer, we get hold of it through feeling and experience.

An adequate urban design should know how to create folds to preserve diversity. Such diversity has always been attached to these stages upon which daily life plays out: residential communities and courtyards, the neglected elevations of high-rise housing. Ten years ago, when Wang Xin and I took pictures outdoors together, he told me: "When people have lived long enough in these buildings, the elevations began to tell stories." What was he referring to? With enough time, the residents' ad hoc additions, the traces of their use, the stacks of household items, their interests, all begin to appear on the balcony, making it a public-facing stage for private life. These elevations look like cabinets of curiosities, recording and displaying decades' worth of home lives. In this way, the balcony deserves to be the *stage of life* in name and in fact.

We see and we imagine, therefore the city is. It reminds me of Le Corbusier's *megastructure*: under miles of elevated road there are 6-meter-high floors in a stepped distribution. People build houses within the space as if making an artificial mountain city. As for the steel-and-glass skyscrapers of the Lake Shore Drive Apartments, there won't be any rich landscape of life on its hard and smooth reflective glass curtain walls. In my opinion, Guangzhou's *Super Wenheyou* is not a spectacle but a legitimate demonstration of the natural urban landscape in East Asia. Once it disappears, it is quite difficult to reproduce such a landscape. This urban design is closer to the humanist ideal of settlement than any Western urban design has accomplished.

An elevation in an old housing district behaves like a courtyard on a vertical plane. It is where the landscape of private life exposes itself to the public and the natural urban landscape is staged. Eliminating unpermitted construction and wrapping buildings with cheap plaster and paint is an unwise decision for the renovation of urban environments. It wipes out the spatial details and smooths out the folds of diversity within an urban area. It reduces the traces of life to ash. It's no better than the inexpert restoration of cultural relics.

It would be wonderful if the hard architectural boundaries in our city could be softened, even just a little. We are capable of exploring the invisible parts of cities by sense and imagination. Instead of understanding the city through big data, we should encourage the enrichment of the elevations of urban architecture, which effectively communicate information to the public through sensation alone. This has little to do with form. If there's one place that contains characteristics similar to those mentioned above, it's Manhattan in New York City. The building elevations may be sterile and monotonous, even in the East Village, but life is accessible from the inner court, from the *rear window*. A splendid city is quite similar to the Naples Walter Benjamin described, and this is precisely what a typical Asian city is like. Fascinating deep structures are found in abundance in Shanghai, Hangzhou, Wuhan, Chongqing, Hong Kong, Taipei, Tokyo, Bangkok, Dhaka, Mumbai, and more. They are demonstrating a possible future for cities. Is it necessary to invent a special theory of urban design for Asian cities? Who is qualified, who has the courage and insight to start practicing?

●透明的舞台

金秋野

希区柯克在《后窗》中复现了典型的纽约街坊式住宅的内院场景。人们可以透过窗口互相窥视，真实的生活总比戏剧更引人入胜。然而透过窗口所能窥见的毕竟有限，何况还有窗帘。阳台则让生活更直接暴露在外。严格地说，阳台是私生活的面孔，当身体被衣服覆盖，人总是要把脸露出来，再戴上社交表情的面具。阳台因此是有表情的。《我的天才女友》再现了上世纪50年代那不勒斯的贫困街区，朝向内院的阳台不仅是私人生活的窗口，也成了邻里、尤其是主妇们隔空交流的终端，当绝望的女主人将物品一件件越过阳台扔出来，私生活开始朝着公共空间呕吐，内院回荡着清晰有力的嚎哭，女人们站在各家阳台上面面相觑。阳台成为透明的舞台，上演生活的悲喜剧。

当代的公寓住宅里，阳台代替了窗口，成为私人空间与公共生活的联通器。有阳台的城市住宅，是一种独特的居住模式，它重新定义了城市中内与外的关系。以北京为例，传统的四合院本身可以看作院落式别墅群，围墙和紧闭的大门限定了内外，很多高档私人别墅群其实也在模仿这种硬边界模式，住宅区实际上是没有什么公共性可言的。杂院化的四合院则刚好相反，本来内向的私人空间突然被不同的家庭占据，彼此半敞开，孩子大人随意串门，此时隔绝私生活与公共生活的不是空间而是时间：夜晚关门闭户才能换来片刻宁静的私人领域。相比之下，高密度的城市集合住宅反而兼容了两者，它让公共生活在“小区”里发生，私人生活浓缩为防盗门后不可见的“隐私”，又从阳台露一丝马脚。

但在寒冷的北方，阳台一般是要封闭起来。完全封闭的阳台，跟外窗也有所不同。阳台一般不做保温层，洞口面积也相对较大，虽然归属室内，依然可以被看作“半内半外”的过渡空间。本雅明是这样回忆小时候外婆家的封闭阳台的：

“在这些隐蔽的房间中，对我来说最重要的却是那个内阳台。或许是因为它的家具简陋，很少受到大人们的重视，或许是因为街上嘈杂的声音轻轻传上来，也或许是因为我可以在这儿看到有看门人、儿童以及手摇风琴演奏者的陌生的庭院。其实内阳台向我展现的更多是声音而不是人物，因为这是在一个富人区，这儿的庭院从来不太热闹繁忙。在这儿干活的人也多少沾染了一些他们有钱的主人们般的悠闲，一周中一直余留着星期日的一些气氛，星期日也因此成了内阳台的节日。其他房间似乎都不够密封，不能完全包裹住星期日的气氛，任它如流水般从筛孔中渗漏出去了。只有这个内阳台与那些插着地毯架子、有着其他内阳台的庭院遥遥相望，把星期日紧紧裹住。从十二圣徒教堂和马太教堂传来的沉甸甸的钟声装满了内阳台，每一声回荡都不会从这里漏掉，直到夜晚，它们依然在这里层层叠叠，久久不散。”

对屋内的人来说，阳台是联通外界的窗口，在小孩子的世界里，甚至是独立面对外界的唯一窗口。但它与外界连而不通，甚至于，它处于家庭生活的末端，由外及内而复归于外，而存留的最小尺度的自然。如此说来，你也可以把它想象成一个敞开式的天井或内院，给起居室或客厅的外窗多一层保护，使其不至于直接面对市井人生。

这样，即使是已经高度室内化、功能化了的封闭阳台，也与它的高级形式——露台、门廊、屋顶花园一道，甚至与居士的庭院、文士的庭园、隐士的山林一道，象征着柴米油盐之外的无限的远方。这是家居生活基因中所携带的自反因素。“我想要个大阳台”，其实你想要的是无力征服的高山大海。

七月初有一则新闻：浙江宁波一位房东把阳台隔成房间，造成熟睡的租客凌晨从十楼坠落。之前也听说过将阳台改造成卫生间，导致阳台整体脱落的。这些大概是阳台改造的极端案例了吧！正常的家居生活中，阳台依然承载着某些半室外功能，比如儿童活动室、健身房、温室或小型菜园、动物饲养场或小型水族馆、洗衣间晾衣间，甚至储藏室，这些原本应当由庭院来承担的功能，不得不塞在一米多宽的过渡空间中。阳台因为塞进了这么多额外的功能而呈现出不同的形象特征，因人的使用而生动多样。

也可以像我在“小大宅”中那样，通过室内空间室外化的手段，将阳台处理为茶室或会客厅，感觉室内空间一下子扩大了许多。但新的问题也随之出现：本属于外围服务性功能的阳台，一下子变成居室的重点乃至核心，必然要与其他空间进行功能置换，比如晾衣服要到卧室或书房，儿童车要放到楼道里，以此类推。对于足够大的房子来说，这都不是问题，小房子就捉襟见肘了。小房子有千般好处，最大的弊端是缺少空间层次感，像孤岛。空间过于局促，没有足够的余裕去感知深度与内外。

总之，如果把住宅看作建筑的终极类型（自古以来建筑的庇护所身份都是第一位的），那阳台的建筑学意义必须附着于住宅之上，它是家庭生活对外显现的一角，因为不定形的内容物而多姿多彩。阳台因向外敞开，同时也是争取额外面积的主战场，很多极限延伸的鸽子笼、花盆架和飘窗，让建筑立面变得异常丰富。我国居民似乎有向外扩张的习惯，与之恰成对照的，是人们对套内空间的巨大浪费和混乱使用。普通城市公寓的阳台面积不超过5平方米，实乃居民最小限度的空间操作。阳台可能无可避免地成为家庭的设备中心，如新风主机、烘干机、电动晾衣杆等都设置在这里，一些卫生洁具、五金工具、甚至不用的电扇、台灯，生病时用过的拐杖和许久不穿的旱冰鞋、装修时剩余的地板和墙漆、舍不得扔掉的绿植和闲置的猫砂盆……建筑师真要三头六臂，才能把这么多大小不一、形态各异的东西做妥善的收纳和隐藏。若非如此，则阳台的“多样性”只能是不得已。

为了避免这种“不得已”带来的视觉杂乱，美国各州有严格的法规，限制公寓阳台上物品的形式和种类。非但如此，还会用黑色的纱网，对阳台栏杆进行遮罩，人为强制“整齐划一”。一些有消防作用的阳台因为连接着垂直消防梯，更是禁止摆放任何物品，甚至都不能在阳台上抽烟——公寓都有固定的吸烟区。出于对邻居的尊重，居民也主动放弃到阳台上去唱歌、弹吉他。功能受到这样的限定之后，阳台的使用效率也大大降低了，人们很少会出现在阳台上，生活形式的缺乏，也让阳台成了虚假的“生活景观”。

那么，什么是真实的“生活景观”呢？

很多亚洲城市至今保留着某种“生态群落”的特征，多年沉积的生活痕迹，让住宅区充满了错综的物品和冗杂的信息，这些内容未必都是美好或体面的，但正如池塘里也会有腐草、淤泥和动物的尸体，正常的城市生活也会有正背两面。如果一个城市只是对外呈现正面而压制背面，那么生态系统本身是不完整的。在美国多数的郊区化住宅中，不仅不让背面显露，甚至希望将其铲除，结果家家户户平整的草坪，就真的成了“景观”而不是“生活”，事物的表里分割开来了。美国的中产阶级住宅区几乎是没有街巷生活的，人们跑步或牵着宠物匆匆而过，见不到老人和儿童，一天中的大部分时间，街道都是一片死寂。一切都美好而虚假，在我眼里那不是真实。“生活景观”就是生活方方面面的完整呈现，它必须是兼收并蓄的、相反相成的，正面和背面根本无法分开，有含苞吐蕊也有枯枝败叶，就像树林或草原。新陈代谢是生命的法则，没什么好羞耻的。

“景观”这个词变得越来越陌生，大概与它的“数字化”发展同步，城市、大地、基础设施，当这些一眼望不到边的事物成为景观思维的对象，景观就成为“观念”而不是个体肉身感知的对象。说到底，景观也有个尺度问题。对我而言，不能靠身体去感知、不能用我的直观认知去了解的事，必须靠闭上双眼、调动意念来模糊把握的事，都可以归为“概念”行列。这样会不会太个人化了呢？我有一个假设：地球上60亿人，身体和知觉的构造都差不多一样，我个人的感受因此具有某种普遍性。这个假设跟时下流行的假设不同，在“大数据”这个流行词背后潜藏着一个总体观念：个体感觉既不真实也不可信，必须对事物有更大尺度的、更准确的定量认知，才能服务于未来的人居环境设计。

我对“概念”有本能的怀疑。我所能看见的、我的身体认为真实的景观，是大树底下的躺椅、山巅可供人登临的大石头、胡同里层层叠叠的花盆。再远的远景，靠眼睛而不是身体才能体会的东西，在我看来等同于贴图。当我想象生活景观的时候，我想到的是什么呢？大概描述一下：应该是无穷多的胡同小院子、街心小公园和家家户户的阳台，无限地绵延下去，缩小在图纸上，没有美丽的拓扑形，没有大地艺术的震撼，只有细密和琐碎，无穷的褶皱里蜷缩着无数个人生，有秩序也有混乱，正面和背面叠加在一起，难分彼此。这就是“生活景观”。

生活不是戏剧，不是做给谁看的。如果它恰好可以被人看见，通过它的物质外壳——房子、阳台、小院子或街心花园，而不是大马路、行道树或立交桥，那它就同时具备了戏剧性的一面，获得了舞台的性质。这种生活的直观呈现是美的，正如青草池塘是美的。高山大海之所以美，是因为连续表面上分布的无数的自然生命。否则拓扑地形有什么价值？美是不可以通过概念来创造的。极多的个体创造了极高的多样性，它们凑合在一起成为极高信息量的空间环境，这就是我们的城市。城市是大数据吗？是的，远高于我们今天认为具有重要意义的大数据。

只是我们可以不用硬盘存储它，而像作家一样，通过感知和想象来把握它。

如果一种城市设计能创造褶皱、保存多样性，那它就对了，否则就错了。城市中的老旧小区或大院、疏于管理的高层住宅的立面，显现着这样的多样性，是生活的舞台。十年前我与王欣去拍照的路上，他对我说："你看这些老楼房住得久了，立面就变透明了。"这是什么意思呢？住得久了，又没有被严格地清理过，阳台上就都是功能、都是使用的痕迹、都是物品的堆叠、都是主人的趣味，是私人生活向公共空间的展示。立面成了生活的多宝格，仿佛透过它能看到数十年生活的样貌，坚如磐石。这个意义上讲，果真是"透明"的。

半靠视觉，半靠想象，城市在我们的内心绵延。我想起勒·柯布西耶的"陆上巨构"，绵延数十公里的高架路，下面是阶梯状分布的6米高的楼板，人们在空间造房子，好像人造的山城。像湖滨公寓那种玻璃和钢铁的居住摩天楼，坚硬反光的平滑边界让它不再有丰富的生活景观可言。广州的"超级文和友"之所以具有示范性，不是因为它创造了奇观，而是因为它提示我们重视东亚城市充满烟火气的自然风貌，一旦推平再难复现，它比任何西方城市设计的经典案例都更接近于人类聚居的人文理想。

老旧小区的立面有点像垂直的杂院，本来不可见的生活景观因此呈现在人们眼前，把平铺在地表的烟火气给叠放在空中，舞台化了。为什么治理穿墙打洞的结果是城市环境的劣化呢？因为它不仅抹平了城市空间的细节，也删除了空间的褶皱，用劣质的灰泥和涂料掩盖了生活的痕迹，其效果跟拙劣的文物修复有何区别？

如果我们的城市的硬边界能稍微溶解一点就好了。不需要很多，一点就可以。我相信人都有充分的感受力和想象力，能够由表及里地探索城市不可见的部分。我们的目标不应该是用大数据去把握这个城市，而应该在城市界面做文章，提高它可体验、可感知的有效信息量。这件事跟形态的关系都不太大。美国的城市中，似乎只有曼哈顿具有类似透明的属性，但反映到建筑立面，即使东村，正面也依然是单调荒芜的，想要一窥人们的生活世界，只能走进内院，仰视"后窗"。美好的城市应该像本雅明笔下的那不勒斯吧！可那不就是亚洲城市的常态吗？

翻翻老照片，上海、杭州、武汉、重庆、香港、台北、东京、曼谷、达卡、孟买，各个都有摄人心魂的深度结构，预示着未来城市的种种可能。是否有必要发明一种根植于亚洲的城市设计理论？谁又有这样的勇气和洞察力呢？

Balcony 透明的 137
A Stage of Life 舞台

Blossom
朵
Apartment
楼房花
138
Blossom
朵

Apartment Blossom 楼房花朵 139

Apartment Blossom 楼房花朵 139

Apartment
Blossom
楼房花
朵
141

Apartment
Blossom
楼房花
朵
142

Blossom

Apartment Blossom

楼房花朵

Apartment
Blossom
楼房花
朵
145

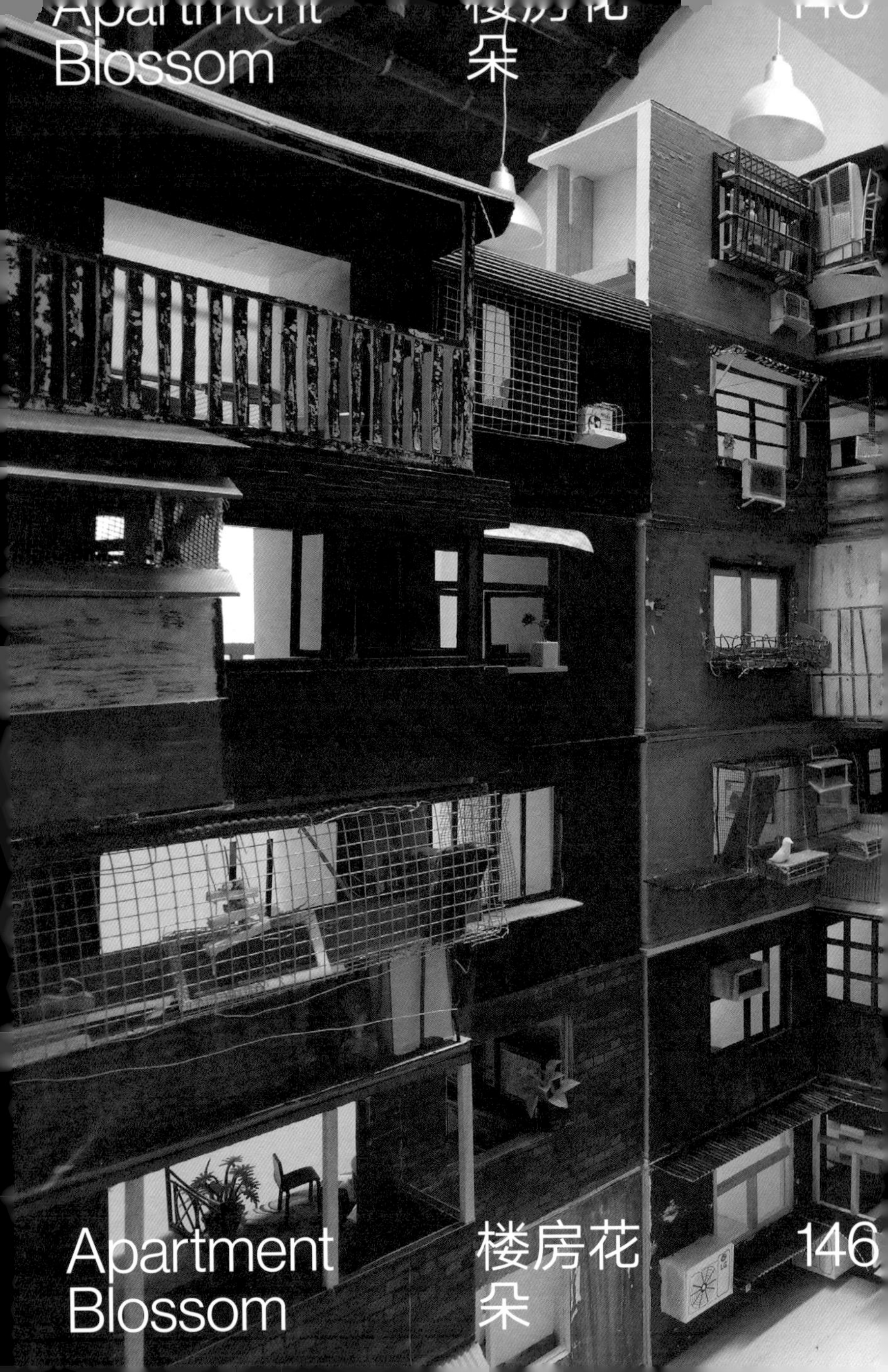
Apartment
Blossom
楼房花
朵
146

Apartment Blossom

楼房花朵

147

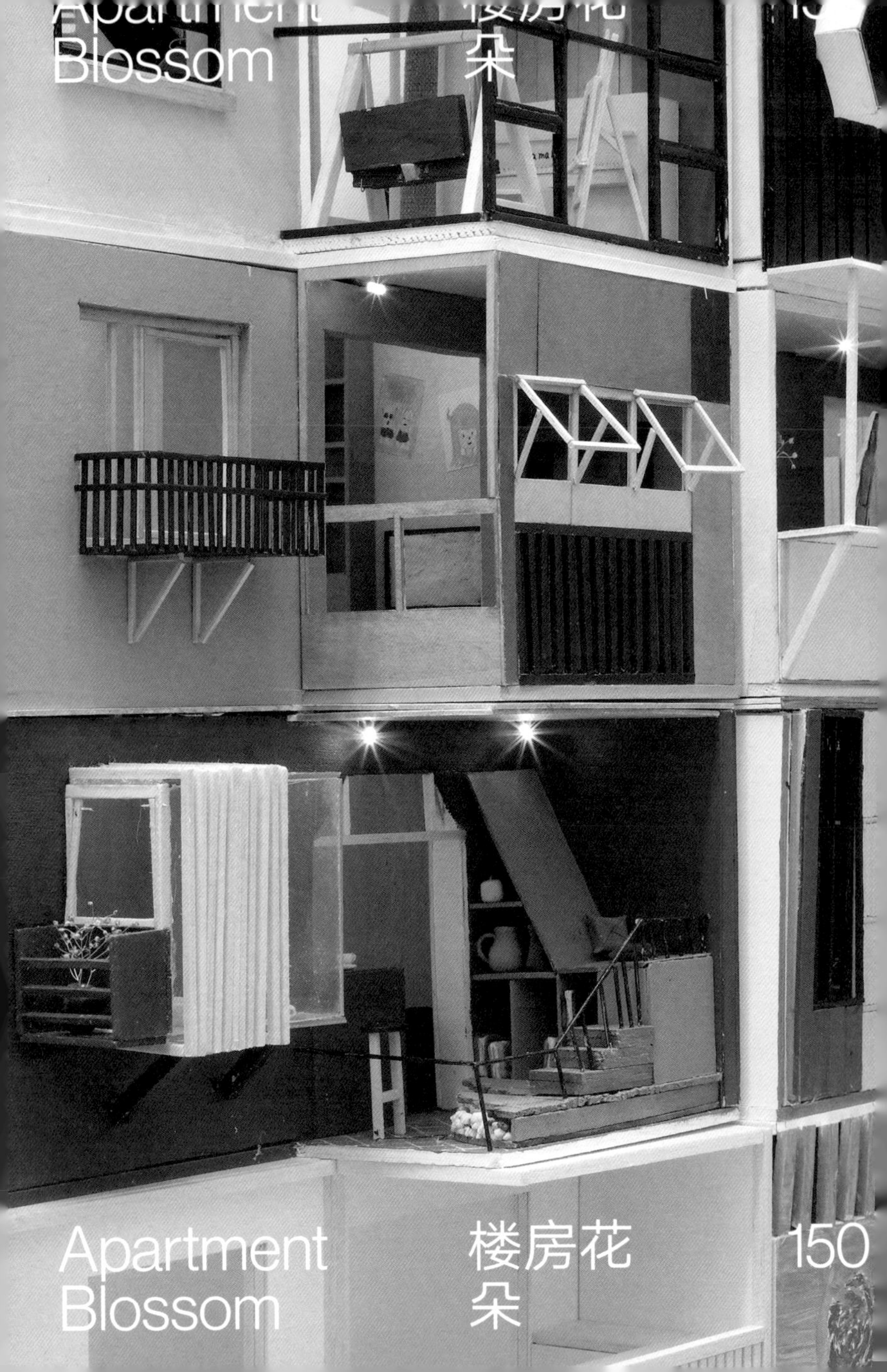
Blossom
朵
Apartment
Blossom
楼房花
朵
150

Apartment
Blossom
楼房花
朵

Apartment
Blossom
楼房花
朵

楼房花
朵
152

秋水
画室
19325645842

Blossom 朵

Apartment Blossom 楼房花朵 154

Blossom 朵

Apartment
Blossom

Apartment
Blossom

楼房花朵

Apartment
Blossom
楼房花
朵
158

Apartment Blossom
楼房花朵
159

Apartment
Blossom
楼房花
朵
160
Apartment
Blossom
楼房花
朵
160

Apartment Blossom
楼房花朵
162

Apartment
Blossom
楼房花
朵
164

Apartment
Blossom
楼房花
朵
165

Apartment
Blossom
167

Apartment Blossom

楼房花朵

168

Apartment Blossom

楼房花朵

168

Apartment Blossom 楼房花朵 169

Researchers

Tang Xinran
Zhao Huiwen
Wang JingChun
Zhou Wenxuan
Li Jinxi
Chen Yifan
Yan Jiarong
Xu Xinze
Zhang Yuntao
Che Yuzhe
Zhang Guohao
Zhang Yichen
Dong Jinyan
Luo Jiao
Wu Lingxi
Zhang Fuhan
Jiang Chaoyang
Chen Yichun
Jiang Sikui
Gan Rui
Jiang Xin
Zhao Minyu
Cui Miaojie
Jin Lingxi
Su Shuang
Wang Yiyi
Wang Yitong
Xue Chen
Shi Boxian
Liu Wenbo
Liu Tingzhou
Yan Xin
Wang Zirui
Li Jian
Tian Yuhang
Jia Xinyu
Zhang Yuchen
Deng Binbin
Miao Zuyang
Meng Zhaoqi
Niu Jiahui
Liu Mingqing
Yang Hongjie
Wei Jiaxin
Meng Yufan
Jia Xiaoxuan
Zhao Xintian
Qu Aojia
Wang Xinyu
Bai Yufei
Li Yiran
Yu Zhaofeng
Sun Peixu
Ma Renjie
Li Muyun
Jia Yunxuan
Han Hongzhe
Ding Yun
Li Chunyi
Ma Weiwei
Yang Haodi
Zeng Zihao
Zhang Jiani
Li Yu
Jiang Zhihan
Ke Tianshu
Ma Kaicheng
Wang Ding
Li Jiayi
He Yue
Shen Jiahui
Li Jiaming
Man Yi
Sui Mujie
He Qing
Yao Anqi
Luo Junjiang
Wang Haocheng
Lu Le
He Lutong
Yang Yiming
Wang Yukai
Huo Guangda
Wang Yuxuan
Kang Tianjiao
Zhang Ziyu
Yang Qixiang
Huang Yulan
Yang Helin
Hou Zhanmin
Yao Siyun
Ji Mu
Liu Yuxin
Huang Ranran
Meng Yuxin
Pan Xijie
Lai Xianzhuo
Sun Lei
Zhang Qiuyan
Zhang Aining
Zhang Ruiyi
Wang Yihan
Zhang Kaiyun
Li Hang
Chen Hongyu
Yang Yuxuan
Yang Jinyi
Han Yiyang
Li Jingchun
Chai Ziyan
Zhang Lingyu
Xu Jiacheng
Guo Jiaxu
Lin Suncheng
Sun Mingjing
Zhang Zhenyi
Wu Yunuo
Mi Jiayi
He Yiran
Liu Yutong
Liu Quanzhi
Wang Ziyue
Gao Siyu
Liu Bingxuan
Wang Jiaqi
Chen Xinran
Du Ruofei
Li Jiani
Tao Lin
Liu Yiran
Zhu Chen
Lou Ying
Wu Yufei
Liu Chenghua
Wang Peidong
Sha Yuheng
Jiang Jiaxing
Liu Haoran
Zhao Leqing
Hu Wanning
Li Hongying
Liu Xinyao
Song Yunge
Zhao Wenyi
Wu Yiran
Du Siyu
Deng Xinyue
Yang Hui
Pu Yi
Lin Menglin
Yang Zixin
Li Zechen
Li Wenxiu
Ding Shuo

Wang Chenyu
Zhao Xinliang
Cao Xiaoran
Wu Yihan
Zheng Yan
Song Qiyuan
He Yuheng
Huang Yi
Zhang Hongli
Niu Zichen
Zhang Chenran
Yu Lufang
Yang Wendi
Wang Zhanyu
Cui Jingya
Chen Zonghao
Wu Xilin
Chen Yuting
Jiao Dongjiang
Li Zhaokang
Lu Yingzhi
Luo Ziyu
Hao Tianxiao
Tuo Yunfu
Zhen Shi
Wang Xiaoge
Han Libin
Huang Xinyi
Mi Xinyu
Jia Damu
Qi Zihan
Fang Jiyuan
Ning Yang Hongwei
Jin Yuqing
Wu Mengdi
Li Difan
Jiang Xinyue
He Zhongyi
Luo Zhengzheng
Hu Hang
Xi Jingyu
Wang Hansheng
Liu Kaihang
Zhang Haosu
Wang Ziyu
Li Wen
Zheng Qingyuan
He Peiyi
Xue Caiying
Chen ge
He Yue
Liu Peikai

Research Directors

Jin Qiuye
Li Han

Instructors

Liu Ye
Cheng Yanchun
Jie Xiaofeng
Xu Zheng
Meng Fanlei
Li Luyang
Zhou Yi
Wang Yao
Sun Li
Yang Zhen
Chen Zhiduan
Zhang Zhenwei
Sun Qiaoyun
He Ding
Wang Bing
Teng Xuerong
Zhang Yu

Photographer

Zhang Xintong

Translators

Li Mengzhou
Andrew Chittenden
Cui Zhongding
Travis Du

研究人员

汤欣然
赵卉文
王静淳
周文璇
李锦曦
陈一凡
严佳容
徐鑫泽
张云涛
车禹喆
张国灏
张一尘
董津言
罗蛟
吴泠茜
张复涵
姜朝阳
陈奕纯
蒋思奎
甘锐
蒋昕
赵玟瑜
崔妙捷
金泠希
苏爽
王懿亿
王一桐
薛忱
史博先
刘文博
柳汀洲
闫鑫
王子睿
李健
田宇航
贾昕宇
张宇辰
邓斌彬
苗足阳
孟昭祺
牛佳荟
刘明庆
杨宏杰
魏佳欣
孟宇凡
贾晓轩
赵馨恬
瞿傲佳
王心语
白雨霏
李一然
于肇锋
孙沛煦
马仁杰
李牧云
贾云轩
韩鸿哲
丁云
李淳毅
马微微
杨皓迪
曾籽豪
张嘉倪
李宇
姜智瀚
柯天舒
马楷承
王鼎
李佳仪
贺悦
沈佳慧
李佳茗
满奕
隋沐杰
何晴
姚安祺
罗俊江
王浩程
芦乐
何陆童
杨逸茗
王昱凯
霍光大
王宇轩
康天骄
张梓钰
杨启祥
黄玉兰
杨和霖
侯占民
姚思芸
基木
刘雨欣
黄冉冉
孟雨欣
潘熙婕
来显卓
孙蕾
张秋妍
张艾凝
张睿漪
王祎晗
张锎匀
李航
陈泓宇
杨宇轩
杨金益
韩伊洋
李敬淳
柴梓艳
张凌雨
许嘉诚
郭佳旭
林孙诚
孙铭婧
张镇屹
吴雨诺
米嘉仪
贺怡然
刘宇彤
刘泉志
王子悦
高思雨
刘冰萱
王嘉琦
陈欣然
杜若菲
李佳妮
陶琳
刘伊然
朱辰
楼颖
吴雨霏
刘成华
王沛栋
沙雨恒
蒋家星
刘浩然
赵乐晴
胡婉宁
李泓滢
刘昕瑶
宋芸歌
赵文宜
吴怡然
杜思雨
邓欣悦
杨惠
蒲艺
蔺梦琳
杨子昕
李泽晨
李文秀
丁硕

王晨宇
赵欣亮
曹笑然
吴逸寒
郑妍
宋奇原
何羽珩
黄祎
张宏俐
牛子辰
张辰冉
于璐方
杨雯迪
王湛宇
崔净雅
陈宗浩
吴禧霖
陈昱廷
矫东江
李兆康
卢映知
罗子渝
郝天啸
拓云福
甄实
王晓格
韩立斌
黄心怡
米馨雨
加达姆
齐紫涵
方纪圆
宁杨弘威
金雨晴
吴梦迪
李迪凡
姜欣玥
何中仪
罗争争

扈航
奚景煜
王瀚笙
刘凯航
张昊苏
王子郁
李文
郑清源
和沛怡
薛才莹
陈歌
何阅
刘佩岂

研究主持

金秋野
李涵

指导教师

刘烨
程艳春
揭小凤
许政
孟璠磊
李路阳
周仪
王垚
孙立
杨震
陈志端
张振威
孙乔昀
贺鼎
王兵
滕学荣
张羽

摄影

张欣桐

翻译

李蒙洲
安德鲁·奇滕登
崔中鼎
杜大有

Li Han, principal of Drawing Architecture Studio, National Class 1 Registered Architect in China. He received B. Arch. from Central Academy of Fine Arts in Beijing China and M. Arch. from RMIT University in Melbourne Australia. His practices include architectural design, architectural drawing and urban studies. He is the author of graphic novels *A Little Bit of Beijing* and *A Little Bit of Beijing·Dashilar* and the translator for the Chinese version of *Atlas of Novel Tectonics*. He was the Overall and Digital Category winners of 2018 WAF Architecture Drawing Prize and won the Second Place at 2016 RIBA Journal Eye Line Drawing Competition. His works were shown in many exhibitions, including the Chinese Pavilion and Japan Pavilion of the 16th Venice Architecture Biennale, The 7th Shenzhen / Hong Kong Bi-City Biennale of Urbanism / Architecture, and Architecture in Comic-Strip Form in Oslo, and are in the permanent collections of San Francisco Museum of Modern Art and Shenzhen Pingshan Art Museum.

Jin Qiuye, Ph.D of Tsinghua University; MIT Research Scholar; member of The Architectural Society of China; Professor, Executive Vice-Dean of School of Architecture and Urban Planning, Beijing University of Civil Engineering and Architecture (BUCEA); architectural scholar and critic. Host of Architectural Criticism Research Institute of BUCEA. Important promoters of traditional culture in Chinese Architecture. Research fields include "the translation of the Chinese traditional design language into Modern Architecture" and "the contemporary architectural thoughts and criticism in China". He published 133 academic paper in the past 10 years and is the author of the monograph *Utopia on the Drawing Board* and *Foreign Body Sensation*. He is in charge of the editorial work of serial publications such as *Arcadia: Painting and Garden and Readers on Chinese Architecture and Criticism*. He is also the Chinese translator of more than 30 architectural theory books including *Radiate City* of Le Corbusier and *Transparency of Colin Rowe*.

Li Han and Jin Qiuye are the co-authors of *Hutong Mushroom*, the first volume of *Urban Studies Degree Zero Series* published in 2018.